The Care and Repair of Marine Gasoline Engines

Loris Goring

The Care and Repair of Marine Gasoline Engines

International Marine Publishing Company
Camden, Maine

Published in the United States of America by
International Marine Publishing Company,
Camden, Maine, 1981

First published by Adlard Coles Limited,
1981

ISBN 0–87742–139–0

Library of Congress Catalog Card No. 80–84623

Printed in Great Britain

Contents

List of Illustrations

List of Tables

To Robert and his narrow boats, Martin and his Mercantile versions, and Carl with his submarines. All my nephews who have come under the evil influence of their wicked Uncle and his enthusiasm for boats of any kind.

Preface

I cannot disguise the fact that internal combustion engines of any kind have always given me a great deal of enjoyment. As a small boy I tinkered with a 1·3cc model aero engine that regularly delivered sparks from the coil to send me hopping from the kitchen table. My present ownership of several petrol and marine diesel units involves a lot of pleasurable maintenance, and I hope that this book will convey some of the satisfaction to be gained from keeping engines in good fettle.

LORIS GORING
Brixham 1981

Acknowledgements

I should like to thank all the engine manufacturers whose products I have featured and who have been most patient with my correspondence and questioning. In particular, I would like to thank BMW Marine for their help when I was given access to every part of their marine engine plant and to the test craft on the beautiful Starnberger-See of Bavaria. I am indebted to Messrs Wessman, Wurst, Nilson and Schutze whose expert knowledge of high performance marine petrol engines must be second to none.

Finally, my thanks to all the tool makers mentioned in the book as, without first class tools, working on any mechanical object can be sheer torture. I am fortunate in having quality tools from such people as James Hugh Neill who take as much pride in tool production as the engine makers do in their products.

General Principles of Maintenance

one

My aim in this book is to give a better understanding of marine petrol engines. Every boat owner should be prepared to tackle simple maintenance and, as knowledge and confidence grow, to take on more complex servicing and repair jobs. Apart from the economics of this do-it-yourself approach, there is the safety angle. Too many fools put to sea in ill-maintained craft as the rescue services know to their cost. Even if your boat is well cared for there may still be an unforeseeable emergency. The responsible skipper who knows the engine and can pin-point and repair the trouble may prevent distress and disaster.

More Care Means Less Repair

Care implies sympathy and communication. And how on earth do you have sympathy with a lump of metal? An easy way is to remember how much it will cost you if you don't. Metal has an inbuilt wish to destroy itself. Neglected, it will corrode and rust. Friction will wear moving parts, reducing them to a metal sludge in the oil. The unsympathetic yachtsman shrugs his shoulders and lets it happen – to his eventual cost.

And communication? I don't mean actually chatting to the engine, though I have heard people swear at them when things have gone wrong. Eyes, ears and noses are part of communication, too. Watch out for oil leaks, spots of rust, a smoky exhaust, a fraying wire. Listen as you start up the engine. Is the starter pawl engaging smoothly? Is the battery a bit reluctant to turn over the engine? How quickly is the motor firing up all cylinders? Sniff out danger signals – hot PVC, hot paint, hot oil. Touch also has a part to play. Know which parts of the engine normally run hot, warm or cool, and check them regularly.

Run your own early warning system and minimise the chances of a major – and expensive – breakdown.

Corrosion

The problems caused by corrosion should have top priority. You should study and understand this complex subject. Research into corrosion is going on all the time as its effects cost a fortune. Corrosion eats away at ships, bridges, cars; in fact most metal items in use today. And that cost is not only in money but in precious energy.

Stated simply, metals – and that includes the marine engine – constantly try to revert to their original oxides. Iron, for example, will revert to iron oxide, i.e. rust. The simple chemical reaction created by iron combining with oxygen is aided by the higher temperatures created in the engine. The water used to cool the engine speeds up this process of chemical decay, and the constituents of the water are also an important factor. Most water contains pollutants, which include

Metal	Steady potential, volts (negative to saturated calomel)	
Zinc	1·03	
Aluminium 3003	0·94	
Carbon Steel	0·61	
Gray Iron	0·61	
Type 304 stainless steel (active) 18/8	0·53	
Copper	0·36	
Admiralty brass	0·29	
70/80 copper-nickel (0·47% Fe)	0·25	
Nickel 200	0·20	INCREASING
Type 316 stainless steel (active) 18/10/3	0·18	NOBILITY
Inconel* alloy 600	0·17	
Titanium	0·15	
Silver	0·13	
Type 304 stainless steel (passive) 18/8	0·08	
Monel* alloy 400	0·08	
Type 316 stainless steel (passive) 18/10/3	0·05	

*Inco Registered Trademark

Table 1. Galvanic series in sea-water at 25°C (77°F), flowing at 13 ft/sec (4 m/sec)

suspended solids of various kinds – metal salts, biological organisms and dissolved and suspended gases such as oxygen, ammonia and hydrogen sulphide, a formidable battalion of corrosive elements. Engine manufacturers do what they can to minimise the effects, but you must be on your guard, ready to replace and repair the ravages of corrosion. Each form of corrosive attack will be dealt with specifically in later chapters but here is a general list:

Galvanic or bi-metallic corrosion. When two dissimilar metals are immersed in an electrolyte (sea water is one), an electrical potential is generated between them. An example of the reaction of metals is given in Table 1.

The disastrous effect of galvanic corrosion can be seen in Fig 1. Here, copper pipes were used to supply cooling water to an aluminium gear box. The precaution had been taken of connecting the gear box with short stub pipes so that the latter disintegrated before the

Fig 1. The results of galvanic corrosion when copper cooling pipes destroyed the aluminium stub pipes on a gear box. Note the constriction inside the rubber pipe caused by debris sticking to the rubber.

expensive box. Little consolation when the stubs need replacing every other season at considerable cost. Even worse, nothing was known of the fault until the unfortunate owner found the engine room filling with water because the cooling water pipes had disintegrated. The electrical current flows between the metals causing the least noble (anode) to be eaten away while the cathodic metal suffers little or no damage. Cathodic protection has to be provided for dissimilar metals immersed in sea water. For example, small zinc blocks are used in engine cooling systems, on outdrive legs and propeller bosses, on hulls and shafting. Failure to inspect and renew this protection regularly will result in serious damage to the unit, whether engine or outdrive.

Where the cathodic metal is in a small area or quantity compared to the anodic metal, there can be such minimal corrosion as to make the combination quite acceptable. So, stainless steels may be used as small fasteners in aluminium plating or castings and, to relieve local stress and create a stronger mechanical joint, stainless steel inserts may be put into outdrive castings.

Stainless steels are found in a passive and active state. In passive forms an oxidised layer forms on the surface and this protects the metal from further corrosive attack. In its active form the layer is missing and corrosive attack occurs. When oxygen is excluded from a stainless steel surface it is liable to pitting attack. Bear in mind that where there are elevated temperatures, as in the engine cooling system, chemical corrosive attack is always accelerated.

Stray current corrosion. The generator, starter and all the ancillary wiring on the engine can be a source of serious corrosion, if they are not installed and maintained to the highest standards. Cracked insulation, damp surfaces on isolating pads and exposed connections may allow current to escape, and turn a normally noble stainless steel propeller shaft or bronze propeller into a corroded shambles; it has been known for propellers to drop off in these circumstances. Cathodic protection systems will cope in all normal conditions but will fail if there is a major leakage of electrical current.

Impingement attack. This usually follows when cooling water is forced round pipe bends at speeds which enable it to wash off the protective layer of corroded metal debris. New metal is constantly exposed to attack and if the cooling water temperature is high, the process is very rapid indeed.

Marinisation

The purist would say that the true marine petrol engine scarcely exists today. There was a time when it was economically viable to build engines for a small specialised market. These marine engines often had bulky castings as a defence against corrosion and to give a prolonged life in a salt water environment. The builder did not worry about weight when boats were generally of the heavy displacement type and these slow running engines suited them well.

Today the picture has changed. Speed has become more important and engines are uneconomic to produce in small numbers; they have also become technically sophisticated in design and production. Today's marine petrol engines have to be derived from automotive units so that the huge production costs can be absorbed by mass production. The Ford Company, for example, supply basic industrial units which emerge eventually as Mercury marine power units; General Motors also supply to Mercury and the Outboard Motor Corporation (OMC); Watermota are Ford (UK), while some Volvo Penta engines are also based on Ford blocks. Provided this marinisation is done well there is no problem, but beware the back street company that does a poor job, goes broke and leaves you with a decaying engine and no service or spares supply. You buy an engine once but it needs to be serviced for years.

Design

In later chapters specific design problems are dealt with in detail but a few general points can be made now.

The automotive engine has to have plenty of air around it to allow it to breathe and to assist cooling. It is relatively simple to pump cooling water round the cylinder block and through a radiator, where it is cooled to maintain the engine at the correct running temperature. The marine engine needs a much more sophisticated cooling system. For one thing it is usually installed in a confined space. For another it has to have either bolstered castings to cope with corrosion from raw water cooling or, like its automotive counterpart using fresh water cooling, some form of heat exchanger cooled by raw water to keep it at its correct working temperature. Some small marine petrol units do rely on air cooling, but they tend to be noisy and only suitable as auxiliary power units in the lower horsepower range.

Throughout this book I will be concentrating on four-stroke marine engines, as these far outnumber the two-strokes that found favour some years ago; but first a few points about the latter.

Two-stroke engines. These are simple in design, needing no complex gear to get the fuel into the cylinder for burning and out again to exhaust it. As the piston travels up the bore of the cylinder on the first stroke, it causes a partial vacuum in the crankcase below it. At a specific moment, the lower skirt of the piston uncovers an inlet port for the fuel/oil/air mixture to rush into the crankcase to ease the depression there. The first part of the downward motion of the piston closes the inlet port and then puts the fuel mixture in the crankcase under pressure as it descends. Towards the end of the stroke a cylinder bypass port allows the now compressed mixture to escape into the upper part of the cylinder to form the charge. The piston begins the cycle again, inducing further mixture into the crankcase but this time, as it compresses the charge towards top dead centre of the crank's revolution, a spark ignites it, producing pressure which drives the piston down. A second port opposite the transfer port allows the exhaust gases to escape to atmosphere and, in fact, the incoming fuel mixture charge helps drive them out. Simple in theory and a lot easier to maintain than the four-stroke but there are snags. The petrol/oil mixture produces a lot of carbon which tends to block the exhaust port and exhaust pipe. You have to give these engines precisely the right amount of oil in the fuel mixture or they soot up the spark plugs usually at the most inconvenient time and place. Ignition systems have to be kept in tip-top condition and spare plugs should always be carried. There is not

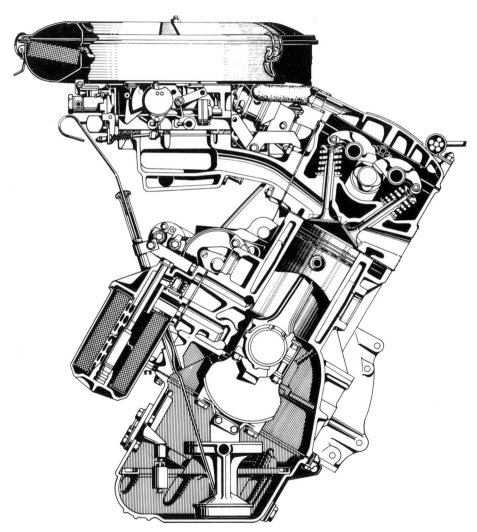

Fig 2. Cross section of the BMW Overhead Camshaft marine engine.

a lot of mechanical work to do but what there is should be done thoroughly. To keep down weight and cost, the small two-stroke is either air or raw water cooled.

Four-stroke engines. These have only lubricating oil below the piston in the sump; all the work happens 'up top'. Starting from top dead centre on the first stroke downwards, the piston begins to create a partial vacuum in the upper cylinder space. A valve (or, in some cases, two valves) opens to allow a fuel/air mixture to be drawn into the cylinder from the carburettor manifold. As the piston passes bottom dead centre, the inlet valve closes and the piston begins to compress the charge. Just before top dead centre a spark at the spark plug ignites the mixture and, as the piston passes top dead centre, the expanding gases from the burn push the piston down. As it gets towards the bottom of its stroke, an exhaust valve (or valves) opens and, with the upward stroke of the piston, the exhaust gases are driven out of the cylinder.

The four-stroke has a lot more moving parts in order to open and close valves, which in themselves need tappets and push rods, cams and springs, making a complex power unit. Although it appears that this cycle has two strokes of the piston which do not contribute to the power, this type of engine is cleaner, generally more reliable, and certainly more economical in fuel than the two-stroke. There is, however, a lot to care for, and let us briefly consider various aspects of maintenance.

Valve gear. Begin by undoing the couple of nuts on the top of the rocker cover to see what is inside. On the majority of engines you will find that the valves are opened and closed, either by push rods operating through rockers from a camshaft low down in the engine block, or by means of a single overhead camshaft as in the high performance marine engine such as BMW (Fig 2). More commonly, push rods (Fig 3) emerge through the cylinder head. Most are of tubular section for lightness and on their upper end a cup accepts the ball of the screw, which is used to adjust the valve rocker clearance. The rockers themselves are mounted on a bar carried on suitable brackets, and are separated by means of light coil springs which reduce noise by taking up end shake.

Referring to the maker's specification for rocker clearance, this is adjusted with a ring spanner, which is used to slacken the locknut. A feeler gauge of the correct thickness is placed between the valve head and the rocker (Fig 4), and the adjusting screw is turned down until the gauge has a nice 'drag-through' feel about it. The difficulty now is to tighten the lockscrew without disturbing the gap. Hold the screw in the correct position with the screwdriver while the nut is tightened down. A thorough check must be made afterwards to make sure all the rockers still have the correct gap.

Fig 5 shows the slightly different design used by BMW on their rockers. Here, instead of a screwed rod to make the adjustment, an

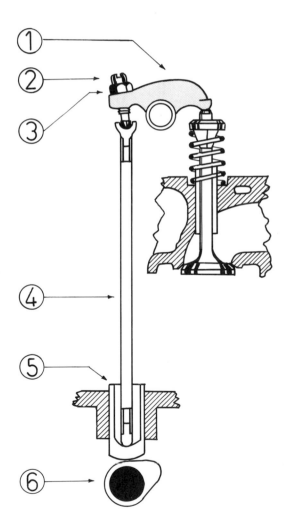

eccentric cam is used. This is gripped between forks on the rocker and is drilled with a hole which allows a special wire tool to turn the cam. Since it is a sideways thrust which tightens the cam, it is easier to keep the gap correct while the nut is being tightened. Care is needed. Do not overtighten as the forks on the rocker can become overstressed.

American manufacturers have developed a design of rocker (Fig 6) which is less expensive to manufacture but just as effective in its function. The rocker (3) is a simple steel pressing, taking its bearing on a spherical spacer, located by self-locking nuts on mounting studs (A), or on a special fulcrum seat (B).

Rockers are occasionally subject to fatigue cracks caused by pressure from the fulcrum point, and groove wear at the pressure point where they push the valve. If such faults are found, complete replacement is necessary.

On many of the engines based on American Ford and GM blocks, there is seldom need for valve clearance adjustments as hydraulic tappets are used to operate the push rods. These highly ingenious devices are very reliable self-adjusting units, worked by oil pressure from the ordinary pressure lubricating circuit. If they seem to be a little noisy, a four-foot length of hose – like a doctor's

Fig 3. Overhead Valve operated by push rod from camshaft in crankcase.
1) Rocker. 2) Slotted rocker adjustment bolt. 3) Lock nut. 4) Push rod. 5) Cam follower; can become grooved and cause clicking noise. 6) Cam on camshaft.

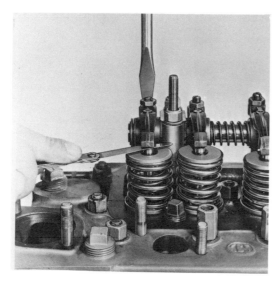

Fig 4. Adjusting overhead valve rockers. The feeler gauge should be a nice sliding fit. A ring spanner is best for tightening down the locking nut while the screwdriver has to be held so that the gap is not disturbed.

Fig 5. Adjusting the forked rocker of the BMW engine.

stethoscope – can be used to listen to them. Place one end near each intake and exhaust valve and put the other end to your ear; hard rapping noises indicate the need for a good cleaning. If the noise is only moderate, it means that small mechanical adjustments are needed. General noise throughout the valve train indicates lack of oil in the lifters, or bad adjustment. Intermittent clicking may mean a minute piece of dirt caught between the ball valve and its seat or, on rare occasions, a flat spot on the valve. Having said that, I don't recommend you to tamper with the valve. The servicing of hydraulic valves, push rods, and rockers is straightforward for the specialist with the right tools, but otherwise might give serious problems. Even

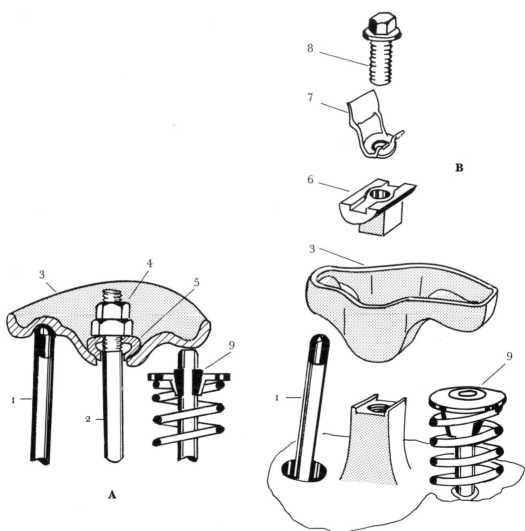

Fig 6. ROCKER DESIGN USING PRESSED ROCKERS

PART A *Section of steel pressing rocker with adjusting nuts.*
PART B *Similar rocker using more sophisticated fulcrum (Based on Ford).*
1) Push rods; professionals check that they are straight. 2) Stud. 3) Pressed rocker. Watch for grooving and fatigue.
4) Adjusting and lock nuts. 5) Spherical fulcrum spacer. 6) Fulcrum seat; base must slot in cylinder head before tightening bolt. 7) Oil deflector. 8) Fulcrum bolt; must be tightened to specification. 9) Valve.

the more common rocker adjustment might best be tackled by the professional. Let me explain.

The gap is there because of the differential expansion of the various parts of the engine cylinder, especially the push rod, the cylinder head, the cylinder head gasket and the main casting. The valves must close and open properly, being fully on or off at their seat at a precise time whether the engine is hot or cold. As the clearance will vary with engine temperature, the adjustment is usually done with the temperature at a minimum. In the case of overhead valves and push rods this would be when the engine is cold, since the rods themselves tend to be comparatively cool, while the upward expansion of the cylinder block tends to increase the clearance, due to the rise in the pivoting point of the rocker. With overhead camshafts the clearance will usually be at a minimum when the engine is hot. Some makers give specifications for two clearances; one when the engine is hot and the other when it is cold. Others only give a 'hot' specification. The outside air temperature has some importance in tappet adjustment. A different adjustment would have to be made on an icy winter's morning than on a warm spring afternoon. I prefer to do the job when the engine is at its normal working temperature and the outside temperature is not extreme. Even then, it is as well to run the engine back up to temperature when you have finished and check all the gaps.

Cylinder head gaskets. These affect the adjustment of the rocker gaps and should be checked and tightened down exactly to the manufacturer's specifications. The cylinder head nuts on a new engine will need attention after about twenty hours running. After that they should be checked once a season or every hundred hours or so. The idea is to keep the gasket under even and specific pressure, so that it does not blow. A blown gasket lets gases from one cylinder escape to another, or to other parts where it certainly is not wanted. Correct tightening down stabilises the distance at the rocker clearance. You need a properly calibrated torque wrench to do this, and you must follow the correct sequence of bolt tightening. The idea of this is to keep the cylinder head perfectly flat in relation to the main casting, so that stress is evenly distributed over the whole surface. Tightening nuts in the wrong order and to the wrong torque may result in a cracked cylinder head – an expensive mistake. When the cylinder head has been tightened down, the rocker gaps must always be re-adjusted.

Sequence for adjusting rocker gaps. The correct adjustment sequence is important. The rocker should be fully off the cam so that the maximum gap is available for adjustment without interference. Each rocker push rod will then be off any high part of the cam in a regular sequence. On in-line, four cylinder engines remember the figure nine. Bearing in mind that No 1 cylinder is at the end of the engine furthest from the gearbox, turn the engine over until No 1 valve is fully depressed by the rocker. It is best to remove the spark plugs and turn the engine over by hand in the normal direction of rotation. Now you can adjust No 8 valve. Continue like that:

No 2 valve open–adjust No 7 (7 + 2 = 9)
No 3 valve open–adjust No 6 (3 + 6 = 9)
No 4 valve open–adjust No 5 (4 + 5 = 9)

and so on, making sure that the total always adds up to nine. Some mechanics put a chalk mark on each adjusted valve.

For in-line six cylinder engines the sequence is the same but the total to remember is thirteen. When No 1 valve is down, check and adjust No 12, and so on. You need to consult the workshop manual for other than in-line engines. On a few old side-valve engines (see Fig 7) the valves have to be removed so that both stem ends and seats can be ground off, a full workshop job necessitating the removal of the cylinder head.

Fire and Safety Precautions

At this point I must stress the dangers of fire and explosion when installation or maintenance is carried out badly. Some authorities, such as the US Coast Guard and UK River Authorities, have stringent rules regarding fire precautions and will not licence boats used in areas under their jurisdiction unless engines are installed and maintained accord-

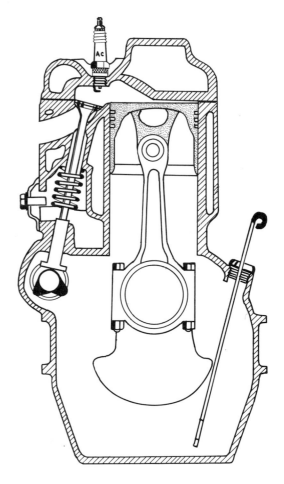

Fig 7. Side valve arrangement of the Volvo Penta MB 10A.

ingly. Despite legislation, boats still catch fire and blow up; all too often it is the human element that creates the risks. Regulations for recreational boats can be obtained from specific bodies, such as the British Waterways Board or Thames Conservancy, and in the USA from the Department of Transportation, US Coast Guard, Washington DC.

The initial responsibility for a safe engine and installation lies, of course, with the manufacturer and boatbuilder but there are many additional factors that are the concern of the boat owner.

Electrical systems. A small spark or overheated wire can set off an explosive mixture of petrol vapour or a puddle of fuel. Don't imagine that because a system is only twelve volts or so there is no danger. Ignition systems, generators, electrical pumps and ventilation fans should all be of a standard which allows them to be used safely in a hazardous atmosphere. All wiring should be of the correct cross section to handle an amperage in excess of that which will cause it to be dangerous, and it must have insulated coating as proof against a surrounding fire. Connections must be well made so that fatigue and vibration will not loosen them. Wires falling off things are a common source of a fire.

Batteries provide not only the current to set off a fire but, whilst on charge, give off hydrogen which is as explosive as petrol vapour. Hydrogen is a light gas which will accumulate under deck heads, so arrange for batteries to sit in a really well-ventilated space, and ensure there is a protective cover a couple of inches above them so that no stray wire, cable or spanner can topple on to them. Incidentally, it is advisable not to wear rings or metal watch straps when working on an engine. Although it is sound practice to have the main battery switches turned off when working, it could happen that a short-circuit across your fingers or wrist might produce a nasty burn.

Fuel systems. Any drip or leak provides a potential source of explosion. If I could insist on one rule it would be: never have a petrol engine installation without a solenoid fuel shut-off valve. These are simple, inexpensive devices that completely close off the fuel supply when the ignition is turned off. I will be dealing with this important subject in greater depth in a later chapter, but would emphasise that, when tuning or testing a petrol engine, you must always be certain to have the backfire flame control in place. Sometimes known as flame arresters, under US law the engine manufacturer is obliged to build these into his engines, and they will be found on every reputable power unit. If a backfire occurs, the resulting flame from the carburettor intake is doused by a metal grill. This works on the principle originally discovered by Humphrey Davey when he invented the miners' safety lamp. The device takes the heat out of the flame so that it is retained within the engine.

Fire Control. This is a very necessary safety aid even in first class operating conditions. There are legal requirements in some areas but the sensible boat owner will provide himself with a tested means of combating fire. Carbon dioxide extinguishers are favoured by some for engine compartment fires as it is clean and effective; both hand and fixed installations are available. My own preference is for BCF (bromochlorodifluoromethane), or in the US the DuPont equivalent known as Freon or bromotrifluoromethane. They are both safe and clean.

BCF and Freon are first class in oil or petrol fires as they are virtually non-toxic until you have had a long exposure – so long, in fact, that there would be little boat left!

I keep a few small hand extinguishers about my own boat as a first aid measure and there is a fixed installation back-up system, by Pyrene, for any major disaster. Anyone working on an engine should have an extinguisher handy.

It seems ridiculous to me that many conscientious owners who spare no expense on fire and vapour detectors and extinguishers, care little about the design and provision of installations to prevent fires developing in the first place. Why buy a cheap engine space fan that will flash up fuel vapour, instead of a gas and spark proof one? How many fuel tank installations have a sensible earth-bonding system to prevent electrostatic ignition and feed stray currents safely to an anode?

I have recently been impressed by a product called Explosafe made by Expamat Explosafe Ltd, a division of the Expanded Metal Company. This aluminium foil is fitted into fuel tanks and drums, and its honeycomb structure dissipates heat so rapidly that dangerous hot spots cannot develop within the tank. It weighs only four ounces per gallon of space and displaces only 0·95 per cent of the volume filled. It would seem to be a product that should be considered for all fuel tanks.

Summary

I hope you are not daunted by this brief introduction to maintenance. Knowledge and confidence are built up slowly, and the first job of any boat owner is to read the Owner's Manual. Learn to identify all parts of the engine. Examine the nuts and bolts and see what they hold together. Check where the water or oil goes and where they can be extracted when necessary. As a complete beginner you may only be able to top up the cooling water, check oil levels and remove and replace spark plugs but, if such simple tasks are done regularly and well, you are already on the way to becoming a good mechanic. This book will help you, but you must also accept that in these days of highly sophisticated power units there will be jobs that only a trained mechanic can do with the aid of special tools – and he will have respect for the **man who knows when to call in the expert.**

two

Service and Maintenance Equipment

Know Your Engine

You will know, of course, the make of power unit on your boat, but each engine has a precise identity in the form of an Engine Identification Plate or an Identification Tag, or both. This plate or tag gives the information required for ordering spares. It is important because in any one year a manufacturer may make a number of modifications to the original specification, perhaps to cut manufacturing costs or improve design or materials.

Sometimes the engine will have only an identification number stamped onto the engine casting. This is sufficient for the manufacturer or his agent to identify the unit.

A copy of the Identification Plate should be made. This is important for ordering spares and also for cases of accident or theft, when the insurance company or the police will have to be informed.

Publications

Useful though I hope this book will be, it is written only to give general guidance and in no way can it be a substitute for the publications issued by engine manufacturers. Many owners ignore the simple servicing procedures laid down in the handbook and then blame the company concerned when things go wrong. Some manuals are printed on grease-and-water-resistant paper, making

them easy to handle with oily fingers. The publications you need include the following: *Owner's Handbook*. This usually covers basic items such as engine identification, operating instructions and servicing. Sometimes it will contain replacement part identification drawings and part numbers. It will probably have a fault-finding guide, basic electrical installation information and an address list of service distributors and suppliers of proprietary equipment. These manuals vary from the excellent to the rather superficial. The quality of service also varies greatly and it is advisable to find out what you may expect. Many companies provide the right replacement parts in the right place at the right time and price. They may also insist on genuine replacement parts being used. This is not merely a matter of keeping the cash flowing but it enables them to give a good service, particularly in quality control.

Small companies may not have such a sophisticated spares service but are often able to offer a personal fast-delivery service direct from the factory. I would recommend using factory-approved spares and service but there are times when a good substitute can save time and frustration. For example, the makers of such items as spark plugs and engine oil filters publish catalogues listing equivalent spares from their own lines. These will have been tested to ensure that they are safe substitutes. If you are not close to a marine depot then there are other sources of spares. If the engine is derived from an industrial unit, spare parts may be available from cement mixer manufacturers, refrigeration and generator equipment makers, heavy goods vehicle or chainsaw depots. Do take care, though, that the spare is the right one; but in an emergency every avenue should be explored. I keep an engine log book with all such sources and part numbers listed.

In the BMW spare parts book, any item common to both car and marine engines has one reference number and any made specifically for marine engines has a second number, making identification easy. When the engine is under guarantee it is foolish to invalidate it by using other than the approved spares. However, it is equally foolish for manufacturers to claim there are no good substitutes for the ones they recommend.

Workshop Manual. Usually this is supplied only to the engine service agents. It gives precise details of the most complex servicing operations—those considered to be beyond the capabilities of the amateur. Even if you are not capable of tackling anything complicated, it is a good idea to ask the manufacturer if you can buy a copy. It will certainly help you to have a better understanding of the engine, and may enable you to detect faults earlier and to give the mechanic a clearer idea of any necessary work.

These manuals are superbly illustrated with exploded drawings, making stripping, cleaning and reassembly of the parts very clear. Quite often specialist tools and expensive service equipment are required, hardly practical purchases for the amateur. Reading the manual makes you appreciate why certain servicing charges are so high, when you realise that this specialised equipment has to be bought by the servicing agent before the engine maker will allow a franchise. The manual will also give all torque values, essential information for the professional mechanic.

Unfortunately for the amateur, some manufacturers are now going in for microfilmed spares and service cards which have to be projected and read in a special optical machine. Cards are more easily up-dated than manuals, but for the amateur the cost of the projector would hardly be justified. Long may the ordinary workshop manual survive. Mercury produce particularly useful and good quality workshop manuals.

Parts Book. Although the Owner's Manual may have a section enabling spares to be identified and ordered, these are by no means universal. If you don't have one, write to the manufacturer to ask if he will supply you with a full parts book. It is a great help to be able to describe exactly what you want when you are at the counter of the service agent and the boat is miles away up a creek. Some of these books contain multi-lingual equivalents.

The engine manufacturer keeps his agents up-to-date and the agent will tell you if there are substitute parts to replace old ones.

Tools

The craftsman rejoices in the quality of the tools he uses long after their price has been forgotten. The bodger buys cheap and has broken tools as well as engines. I make no excuse for urging you to acquire a set of tools that will be a pleasure to use and make the job safer and easier. Using the wrong wrench not only chews up the head of the nut but may slip and damage your knuckles.

Life would be simpler for mechanics if threads and nut sizes were unified. Occasionally, international committees get together to standardise these things, but the people who don't get to the meeting seem to go on in the same old way and even resurrect threads that were made redundant years ago. Modern engines in the USA and UK use American Fine or A/F. Sometimes these wrench sizes are given as American/Unified. For electrical components, British Association or BA is the favourite while power units from the continent of Europe use metric sizes. Whitworth, a thread that was to be made redundant some time ago, still manages to find much use in the marine world, especially where it pops up as British Standard (BS). Table 2 on page 18 gives equivalents for the different systems, but don't expect to keep fasteners in good condition if you use the wrong wrench, even if it appears to be nearly right in size on the table. Only a perfect fit should be acceptable.

Basic Tool Kit

While some fortunate people are able to afford a complete mechanic's tool kit like the Neill set illustrated in Fig 8, others will be happy to build up a set a little at a time. Whichever way, it is a good plan to house expensive tools in a decent cantilever box (A). These are available in both metal and plastic but, if your choice is the latter, make sure it is strong enough as tools are heavy and the box has a hard life. Bear in mind too, when selecting tools for working on boat engines, the inevitable cramped working space.

Screwdrivers (K). Keep a selection to fit any width and type of slot without damaging it and with the correct reach. Short stubby screwdrivers are needed for use in cramped spaces, such as between the forward end of the engine and the bulkhead, while the long ones give a higher torque and reach into deep set places. The electrician's screwdriver, with $\frac{1}{8}$in or $\frac{3}{16}$in blade, should have an insulated handle. You will also have to deal with a variety of cruciform slot screws: Phillips, Pozidriv and the one that has now made Pozidriv redundant, Supadriv. There are definite engineering differences in the three designs of slot, and the correct·screwdriver with the correct size point must be selected if damage to slot and driver is to be avoided. The same design driver is compatible with both Pozidriv and the newer Supadriv heads. The modern cruciform slot is very much less

INS	MM	BA	BS INS	W INS	AF INS	INS	MM	BA	BS INS	W INS	AF INS
0·193		6				0·562					$\frac{9}{16}$
0·197	5					0·590	15				
0·218					$\frac{7}{32}$	0·593					$\frac{19}{32}$
0·220		5				0·600			$\frac{3}{8}$	$\frac{5}{16}$	
0·236	6					0·625					$\frac{5}{8}$
0·248		4				0·629	16				
0·250					$\frac{1}{4}$	0·656					$\frac{21}{32}$
0·276	7					0·669	17				
0·281					$\frac{9}{32}$	0·687					$\frac{11}{16}$
0·282		3				0·708	18				
0·312					$\frac{5}{16}$	0·710			$\frac{7}{16}$	$\frac{3}{8}$	
0·316	8					0·748	19				
0·324		2				0·750					$\frac{3}{4}$
0·338			$\frac{3}{16}$	$\frac{1}{8}$		0·781					$\frac{25}{32}$
0·343					$\frac{11}{32}$	0·787	20				
0·354	9					0·812					$\frac{13}{16}$
0·365		1				0·820			$\frac{1}{2}$	$\frac{7}{16}$	
0·375					$\frac{3}{8}$	0·826	21				
0·394	10					0·866	22				
0·406					$\frac{13}{32}$	0·875					$\frac{7}{8}$
0·413		0	$\frac{7}{32}$			0·905	23				
0·433	11					0·920			$\frac{9}{16}$	$\frac{1}{2}$	
0·437					$\frac{7}{16}$	0·937					$\frac{15}{16}$
0·445			$\frac{1}{4}$	$\frac{3}{16}$		0·944	24				
0·469					$\frac{15}{32}$	0·968					$\frac{31}{32}$
0·472	12					0·984	25				
0·500					$\frac{1}{2}$	1·000					1
0·512	13					1·010			$\frac{5}{8}$	$\frac{9}{16}$	
0·525			$\frac{5}{16}$	$\frac{1}{4}$		1·023	26				
0·551	14					1·062	27				$1\frac{1}{16}$

Table 2. Equivalent for inches, mm, BA, BS, Whitworth and A/F

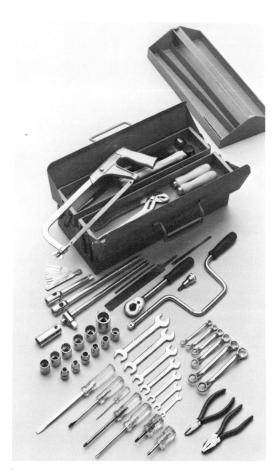

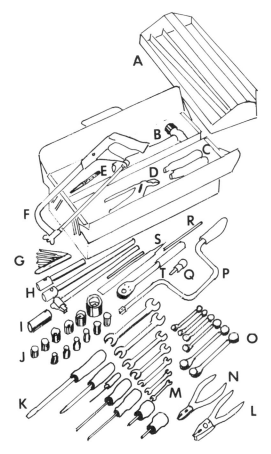

Fig 8. *A basic tool kit selected from Neill Tools.*

A) *Steel Cantilever tool box.*
B) *Engineer's Ball Pein Hammer $\frac{3}{4}$ lb.*
C) *Handles for files.*
D) *Water Pump Pliers.*
E) *$\frac{5}{32}$ Diameter Punch.*
F) *'Eclipse' Hacksaw.*
G) *4 in Feeler Gauge Set.*
H) *'Britool' $\frac{1}{2}$in Bar Handle and extensions.*
I) *'Britool' Spark Plug Sockets $\frac{1}{2}$in square drive.*
J) *'Britool' $\frac{1}{2}$in Sockets to suit your engine.*

K) *Screwdrivers—large medium and small blade types, electricians and 'Pozidriv' types for Pozidriv and Phillips head screws and bolts.*
L) *Pliers.*
M) *Open ended spanners to suit your engine.*
N) *Cutting Nippers.*
O) *Ring Wrenches.*
P) *'Britool' Speed Brace.*
Q) *'Pozidriv' bit with $\frac{1}{2}$in square drive.*
R) *Magneto File.*
S) *8in flat and round files.*
T) *'Britool' Ratchet Handle.*

vulnerable to camming out and is therefore safer to drive, whilst having more power exerted on it. Socket manufacturers such as Neill offer bits for use with their socket wrenches, leading to greater versatility. Remember, a screwdriver is neither a cold chisel nor a lever. Keep the faces of the blade at the correct taper (not sharp) and the blade itself straight.

Pliers (L). Contrary to popular belief, pliers are not universal wrenches. They should be used purely as guillotine cutters without twisting, as this may damage the hardened edge. Specialised pliers are made to deal with internal and external circlips with long nosed versions for holding small items. Waterpump pliers (D) are ideal for reaching into pumps to remove the rubber impeller for the winter. Again, it is best to buy insulated versions for, as well as giving some protection from electrical shock, they are better protected against corrosion, easily cleaned and pleasant to handle.

Wrenches (O and M). Three basic types are available: open jaw (M), ring or box (O), and combination (not shown). Ring wrenches give the better grip distribution, the torque being distributed on each of the six sides of the nut. They are available in six and twelve point openings, and the latter have the advantage of shorter swing in confined spaces. The open jaw types with a different size at either end do save the amateur mechanic some time. Combination wrenches seem attractive until you remember they have equal ends, so you need more of them to cover the full range of fasteners.

Adjustable wrenches are fine so long as they are adjusted correctly onto the nut. A sloppy fit does as much damage as a wrong wrench of any type. An 8in or 10in model will handle the normal run of sizes but it is useful to have a really large one to deal with items such as the crankshaft drive pulley retaining nut.

Socket Wrenches (H, I, J, P, Q, T). Though a good set of socket wrenches means a substantial initial outlay, it is a first class investment as regards speed and safety of working and kindness to mechanical components and fasteners. Neill make a dwarf socket set (Fig 9) with BA sizes from 0 to 8 and AF sizes from $\frac{3}{16}$in to $\frac{7}{16}$in–an ideal set for working on electrical components and smaller mechanical gear. The flexible socket drive gives access in the most difficult situations. Smaller sets have $\frac{1}{4}$in drives but as size progresses $\frac{3}{8}$in, $\frac{1}{2}$in, $\frac{3}{4}$in and 1in drives are available. The general kit should be in $\frac{1}{2}$in. A combination set would be the best buy, with either metric or AF according to the engine, for the main set, backed up with Whitworth. If funds run to it, some sets have, in addition, metric size sockets. Since drives are in universal square sizes, it is advisable to collect items gradually if you cannot afford the high initial outlay. When using sockets there is some danger in not realising the great leverage you are able to exert with long tommy bar drives and the speed drive. I will deal with this point now.

Torque Wrenches. Always tighten down nuts

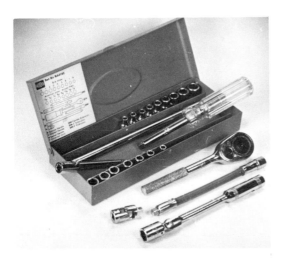

Fig 9. A dwarf socket set by Neill, ideal for those small awkward-to-get-at jobs.

by exactly the right amount. Overtightening puts unnecessary strain on the stud and casting, gaskets will be over-compressed and everything will be that much more difficult to undo next time. Many parts of the engine assembly are critical as regards the tightness of fastenings. The workshop manual will give precise instructions. The safe torque to apply to the nut and stud is directly related to the materials they are made of. Weaker materials, obviously, need less torque than tough ones. On the other hand they should not be under-tightened as if they fall off all hell could be let loose in the engine.

A torque wrench should be considered

Fig 10. Neill EVT torque wrench being used to tighten down the cylinder head nuts on a Watermota marine engine. Precise control of torque applied to nuts on the modern engine is essential.

indispensable for tightening down cylinder heads and all internal fasteners on crankshafts, rockers, connecting rods and the like. I have seen an expensive carburettor wrecked because one side was tightened down so much that it cracked the main casting. These tools are expensive so take great care when choosing them. I have found the EVT series (Fig 10) torque wrenches made by James Neill to be ideal. In particular, the EVTR has a 24 tooth double pawl action which makes it easy to use in a confined space. Another attraction is that it is graduated in all four common torque units of measurement – Newton metres, kilogram-metres, lb/ft, and lb/in. Such versatility is advisable, as the internationally agreed specifications slowly come in.

You will find that cylinder heads may have to be tightened down in several stages. A new engine, for example, takes about twenty hours before the head needs tightening down. This is to ensure that the gasket beds down perfectly and tightening stresses are evenly distributed over each cylinder head stud. Uneven compression of any gasket is going to mean trouble later. A carburettor gasket may leak air, an exhaust manifold noxious gas, or a gasket in the cooling system, water.

Fig 11. Studs being extracted with a Neill stud extracting tool.

Drastic Engineering Surgery

Tools to deal with major engineering problems will be necessary for such jobs as removing a corroded nut with no flats left on it. This can be tackled with a nut splitter. This is placed over the nut; a screw thread is tightened by means of a spanner, and a hardened chisel point splits the nut and takes

it off the thread. It will damage the thread if not used properly.

Studs are often inset into a casting and if they are damaged they must be replaced – easier said than done unless you own a stud extracting tool: (Fig 11). This clamps over the stud and, used with a heavy duty socket drive wrench, the most stubborn stud will be shifted. If a bit of stud is left protruding, it is possible either to file flats on it so that it can be gripped in a wrench, or to cut a slot with a hacksaw and use a screwdriver. If a stud breaks off flush, the best method is to mark the centre with a punch and carefully drill a hole down the stud. Insert a tap extractor tool, hammering it gently into the hole then, as its square section grips the stud, turn it with a wrench and the stud is out.

A stripped thread which would present only a minor worry in a workshop can be a nightmare in a boat. Marine engines are particularly vulnerable due to corrosion problems and the onslaught of strong-armed mechanics wielding a spanner too enthusiastically. For example, brass drain plugs on light alloy gearboxes, especially at salt water drain points, can spell disaster for the threaded section of the casting. Even stainless steel threads into light alloy can sometimes seize solid and the softer metal thread will strip before it can be removed. I will deal with suitable isolating sealants later, but would recommend you to learn about Heli-Coil screw thread inserts made in UK by Armstrong Fastenings Ltd. BMW now use Heli-Coil in their engines and outdrives as a means of stress relief at fastening points. The heat and corrosion resistance offered makes for easier servicing when they have to be dismantled. The procedure is as follows:

Looking at Fig 12:

(A) The damaged thread size is obtained from the fastener. If you are not acquainted with thread types and sizes ask an engineer. Once the thread size is known an appropriate Heli-Coil kit can be bought or hired. This kit consists of variously sized Heli-Coils complete with a matching tap. You will only need a standard drill bit in the size of the Heli-Coil before you can begin. Drill out the remains of the old thread.

(B) Tap the cleaned up hole, using the recommended Heli-Coil tap. Take care to keep it running straight. Parts need only to be dismantled if there is danger of the debris dropping into mechanical sections.

(C) Having loaded the Heli-Coil inserting tool with the appropriate Heli-Coil, it is applied to the tapped hole and inserted by turning the handle until the coil is just below the surface (D). (E) shows a Heli-Coil insert being used for repairing a spark-plug hole thread. There is a coil-removing tool if there is need for renewal. This is unlikely as the engineering accuracy and corrosion resistance should make this a long lasting repair. For blind holes, the tang which drives the Heli-Coil through the inserting tool need not be removed, but for through holes, such as the spark plug, the tang should be removed by bending it up and down with long nosed pliers. Make sure that it does not drop into the cylinder.

In the past engineers tended to drill out damaged threads in a casting and tap out with the next larger size. The system just described enables the same size fastener or plug to be used, with obvious advantages. At the time of writing, Ferralium Alloy is coming into use for the inserts and the properties

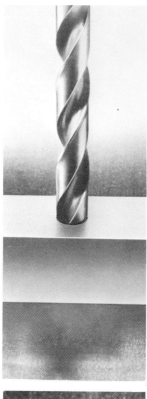

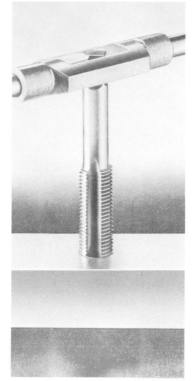

Fig 12A). The Heli-Coil. Drilling out the old thread ready for re-tapping.

Fig 12B). Tapping a new thread to take the Heli-Coil insert.

Fig 12C). Inserting the Heli-Coil with the special tool supplied in the kit.

Fig 12D). Heli-Coil inserted so that it is just below the surface ready to accept a new bolt.

Fig 12E). The Heli-Coil being used to repair a spark plug hole thread.

of this alloy in relation to corrosive environments may be of great interest.

There are occasions when even the largest suitable screwdriver will not budge a corroded bolt. This is a situation commonly found on fasteners into light alloy castings which have been submerged in water, such as on the outdrive housing. The Impack-Driver is the tool you need here. This is virtually a screwdriver that converts a heavy blow with a mallet into torque enough to turn the most stubborn slot head bolt. You must take care that the blow does not damage the casting, but otherwise it works very well.

Gaskets and Sealants

You should keep a small supply of basic gaskets. The water pump will need one each year, and you will soon get to know which other parts have to be dismantled regularly.

It is also a good idea to have some basic jointing and gasketing compounds to hand. The Loctite company manufactures on both sides of the Atlantic and has a good range of products; such as Loctite Superfast Flange Sealant 573 which is used for motor, gear box and outdrive flanges as well as for sealing intricate diecast flanges with fine orifices, such as carburettors. These products are simple to use and are resistant to the solvent action of fuel and engine and gear oils. Mercury Marine recommend Loctite products for use on their engines. In light alloy assemblies,

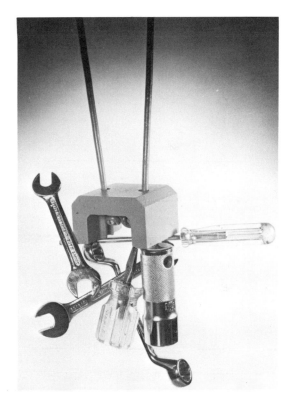

Fig 13. A hefty Eclipse magnet is ideal for retrieving tools lost in bilges and overboard so long as it is kept at a distance from the navigation compass.

I would advise using an active corrosion inhibitor on the fasteners, such as chromate compound to DTD 369. One source of supply is Llewellyn Ryland Ltd, Hayden St, Birmingham 12, England.

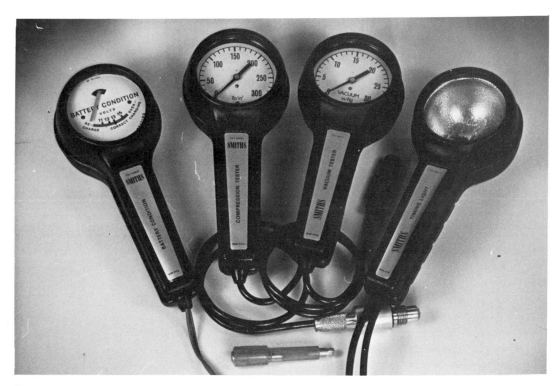

Fig 14. Engine tuning instruments by Smiths. Left to right: Battery Condition, cylinder pressure, manifold vacuum gauge, timing light with HT tester in the foreground.

Tuning Instruments

For the enthusiast, three or four tuning instruments are recommended. They are relatively inexpensive, are simple to use and accurate. Fig 14 shows some of those in the Smiths Industries range, including a battery condition tester, a timing light, a vacuum gauge and a compression tester. The battery condition tester is clipped onto the battery with two crocodile clips and will, after a few seconds, indicate the state of charge. The ignition timing light is necessary for accurate ignition timing, and it will help in determin-

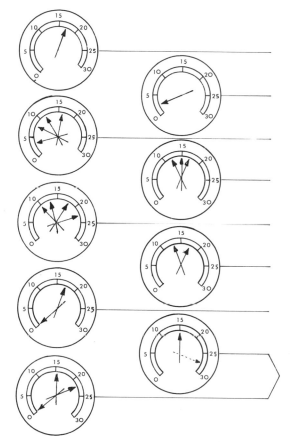

Average and steady *15-21 inches. Confirm good condition by blipping throttle. Vacuum should drop to about 2 then back to about 25 when blipping is done quickly.*

A steady reading below 5 inches indicates a leaking manifold or carburettor gasket. A steady reading between 8 and 14 inches indicates incorrect valve timing.

A reading drifting between 5 and 19 is generally caused by a blown head gasket.

Indicator needle fluctuates very slowly. Check carburettor adjustment, spark plug gap (too narrow) or a sticking valve.

Indicator fluctuates rapidly at idle, steadies as rpm is increased. Valve guides may be worn. Slower swing back and forth on accelerating indicates weak or weakening valve springs.

Indicator continuously fluctuates between low and normal reading at regular intervals indicates a burnt or leaking valve or worn contact breaker.

Indicator drops to zero as engine rpm increased indicates exhaust system restricted.

Indicator holds steady 12 to 16 inches vacuum but drops back to zero when accelerating and climbs to 22 when decelerating indicates worn piston rings. A higher than normal reading (dotted) indicates a choked air cleaner.

Fig 15. Troubleshooting with a vacuum gauge. Most engines will have a normal gauge reading of 15 to 21 inches of vacuum at sea-level. Corrections must be made for elevation above sea level and readings made at normal operating temperature. These diagrams are intended only as a guide.

ing the condition of the distributor and automatic advance and retarding systems. The vacuum gauge is a valuable diagnostic instrument but it does require a ¼in BSF tapping to be made on the induction manifold and, with twin carburettor installations, you need the advice of the engine maker to determine the precise location so that readings will be accurate. Fig 15 gives some advice on the troubleshooting possibilities of the instrument. Using the diagram as a guide, you will soon get used to the normal vacuum readings for the engine and then, with practice, will interpret abnormal readings as they occur. The compression tester will confirm the condition of piston and piston ring.

three

Cooling Systems

Air Cooling

This is the simplest system used on small engines. The air absorbs heat by means of radiation and conduction from the engine's hot surfaces, especially round the cylinder. A fan working in conjunction with cunningly contrived ducting removes it for dissipation into the atmosphere. Maintenance is merely a matter of keeping fan, ducting and engine surfaces clean. Dirt on surfaces reduces radiation and lessens the efficiency of the fan and ducting. Seagull feathers seem to have a great affinity with ducting, and I once saw what was almost a complete nest causing overheating problems on an open launch installation. Air cooling has never been too popular because of the high noise levels. This is due to the nature of the design, mechanical noise being thrown out from the engine direct into the circulating air and from the fan, which not only absorbs a considerable amount of the engine's power but creates its own noise. Much of the noise can be taken care of by good design. If ducting cross sections are generous, the addition of acoustic cladding will absorb some of the noise. Engine manufacturers will advise about practices to provide the correct and necessary volumes of air. Any cladding used must be fire retardant, preferably up to Air Registration Board fire safety specification. Apart from keeping it and its cooling system clean, the air cooled engine requires no other winterisation programme.

The life and efficiency of the air cooled engine is directly related to the quality of the cool air ducted to it. If it is allowed to run in a badly ventilated compartment, the heat dissipated to the walls will pre-heat incoming air. Air entering the carburettor at an elevated temperature will ensure power loss and may also lead to a fuel vapour lock, causing the engine to run irregularly and then stop altogether. Overheating any engine is bad, but it is particularly brutal to an air cooled unit and leads pretty quickly to exhaust valve jamming or seizure of the piston. To reduce these risks to a minimum and give these engines a good life, the following steps should be taken:

(a) Ensure that hot exhaust pipes, especially in 'dry' systems, are well lagged. Some small air cooled engines have a raw water impeller pump (discussed in next section) to give a water cooled exhaust. Lag the pipe up to where the water is injected.

(b) If you are not going to arrange proper ducting and an auxiliary fan to blow cold air in, have a hole in the top of the engine casing to allow hot air to escape. Remembering that cool air is heavy, ventilation should be organised so that air is drawn through bilges to ventilate them. As hot air rises, it can be ducted from the engine to heat a cabin space before being lost overboard.

(c) If you still suspect the engine is running at a higher temperature than is good for it (say on really hot summer days or in a hot climate) consult the engine manufacturer. He may recommend a heavier grade lubricating oil for use in the engine sump. Air cooled engines must have first class ventilation and there is no substitute for this.

Direct or Raw Water Cooling

Water cooled engines have the advantage that the noise from the working of the internal mechanical parts is subdued by the surrounding jacket of water. They are much quieter than air cooled units but more costly to manufacture, because they need a host of items like sea-cocks, water strainers, pumps and thermostats to circulate and control the water flow. All these add to the initial cost and to the amount of maintenance needed. Direct or raw water cooling refers to engines which take water straight from the source on which the boat is floating, which may be fresh water or salt. Unfortunately, all such water, whether fresh or salt, is likely to be contaminated. Pollutants can be a serious source of wear and corrosion within the cooling system. Silt and sand will, by their abrasive action, encourage wear in pumps and blockage of passageways in casting and water galleries. Depending on their chemical composition they will also contribute to corrosion problems. Since marine engines have evolved from industrial and automotive power units where weight has had to be kept down, corrosion problems can become critical. For this reason there has been a decline in the simpler direct

cooling systems in favour of the indirect, where the major part of the engine can still be fresh water cooled as was originally intended. Raw water systems can still give good service life if they are constantly monitored for corrosion, cleaned out while laid up and generally nurtured.

In Fig 16, showing a raw water cooled engine (based on Volvo), there is a sea-cock (B). Volvo have, in the past, also used a combined transom inlet/exhaust fitting. Other power units often draw the raw water supply via an inlet in the outdrive leg. This valve may incorporate a filter for removing weed and other large debris, but this can be replaced with a separate filter in a more accessible place on the engine or bulkhead (D). Filters need regular cleaning. An impeller type pump (often referred to as a 'Jabsco') circulates the raw water supply through the engine and manifold, after which it is injected into the hot exhaust gas stream to be cooled and carried overboard. To prevent the engine casting, cylinders, valves and other working parts becoming over-cooled, the water circulation is split at a distribution housing. The exhaust cooling must have a constant supply of water. The engine needs a supply of water only when it has first warmed the internal water in the casting and head to the correct operating temperature and then only sufficient to maintain it at that temperature. Modern engines have a second circulating pump, usually of the centrifugal type, to ensure an

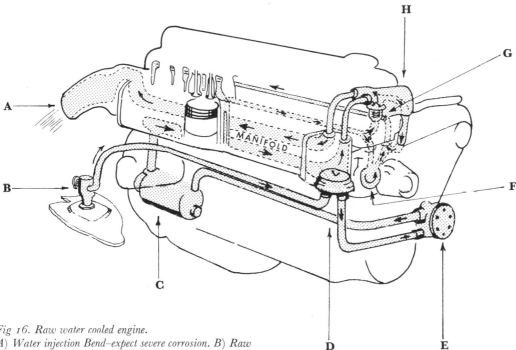

Fig 16. Raw water cooled engine.
A) Water injection Bend–expect severe corrosion. B) Raw
water inlet, here shown as skin fitting; there may be a
transom fitting combining inlet with exhaust or an intake
via the outdrive leg. C) Oil cooler–heat exchange type;
often omitted on smaller engines. D) Raw water filter.
E) Impeller type pump–often referred to as 'Jabsco' type
but that is a trade name; needs servicing–see text.
F) Circulation Pump–often of centrifugal type; large
power units have this extra circulation to ensure even
heat distribution. Smaller engines use only (E).
G) Thermostat (open for this diagram). H) Distribution
Housing; the exhaust must always have water for cooling
the manifold and exhaust when thermostat is closed (engine
cold).

even circulation of water (F), while smaller
and old units have just the single impeller
pump. I will have more to say about the
thermostat (G) later, when the servicing
requirements of the parts are mentioned in
greater depth.

In passing, I would mention that larger
power units may use the raw water side for
ancillary equipment cooling. Thus, raw
water may be circulated through the gear box

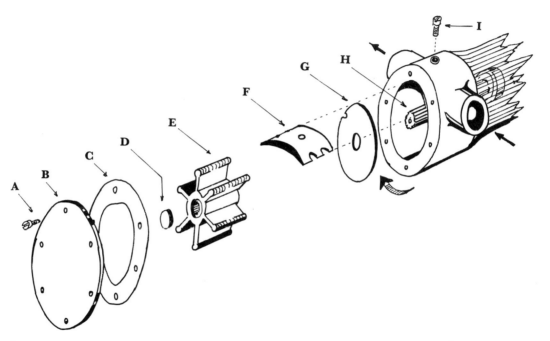

Fig 17. Raw or seawater pump–impeller type. A) cover bolts (6). B) end cover plate. C) paper gasket (on some models 'O' ring). D) rubber plug to cover centre of impeller. E) impeller–must be removed for winter lay-up. F) cam plate or segment. G) inner wear plate. H) splined drive shaft. I) cam plate retaining screw.

to cool the gear box oil before passing on to an oil cooler, and only then to the main engine cooling job. Where this extra cooling work is done, it is even more important to flush the raw side of the system thoroughly at the end of each season and, if possible, at more regular intervals.

The Raw Water Pump. Fig 17 shows a raw water pump of the impeller type. This design is ideal for raw water as it handles pollutants and even small debris without distress, is simple to service and long lasting. They are often known under the trade name of 'Jabsco' in Europe, but many firms make impeller pumps and their working parts are much the same. The main part needing attention is the neoprene impeller. Scouring wears away the end of the lobes as they brush round the pump casing. They age-harden and, if allowed to disintegrate in the cooling system, are appallingly difficult to remove. For a happy life, therefore:

1. Always remove the impeller when the boat is laid up for the winter as constant deflection of the blades ensures that the impeller hardening takes place.

2. As soon as the impeller shows any hardening and will not recover its round shape with the blades radiating straight from the hub or when there is serious wear on the lobes, it must be renewed.

3. Always carry a few paper gaskets as used on the end plate. Lightly grease this gasket when renewing so as to preserve the pump body and cover plate from corrosion. End plate screws and a spare slinger washer, if fitted, are also useful spares. Firms making the pumps often offer a basic service kit.

Seals are used to prevent water in the pump impeller housing from running down the drive shaft. They are long life components but they can start leaking and become a serious source of trouble, especially when direct drive from the main engine casting leaks water to the crankcase. These seals also create an oxygen free area on stainless steel drive shafts and this leads to pitting attack which will, in turn, chew up the seals. If it is a straightforward job to remove the pump and take out the drive shaft for the winter, this should be done, as exposure to the air will help renew the oxidised surface, making it passive. Sometimes impeller pump castings

have a cut-away in the drive shaft housing, which allows a rubber washer on the drive shaft to sling any drips of water away from the pump. A final seal on the engine side is an extra aid in preventing water damage. Check that the slinger cut-out is clear of dust and debris and that the slinger washer is running free with the shaft.

Apart from general wear, real damage is done to an impeller pump when for some reason its supply of water dries up. Maybe when running aground the intake becomes blocked with mud; or someone forgets to open the sea-cock; or, at laying-up time, the engine is run without a hosed water supply connected. The heat generated by friction soon makes the impeller disintegrate, burning off the lobes in next to no time. Never let the pump run without water – and check to see whether the power unit has more than one. Most engines do only have one raw water pump of this type, but the MerCruiser 1 outdrive unit has two: one on the engine and one on the drive unit. It is sensible to apply a little Marfak 2HD grease inside the pump to give initial lubrication when starting up a dry pump after the winter. A touch of glycerine will do as well – and I mean just a touch.

Thermostats are found on both direct (raw water) and indirect (fresh water) cooling systems. Their function is to control the engine's working temperature so that fuel is correctly vaporised thus allowing all parts to run at the right temperature. The problem

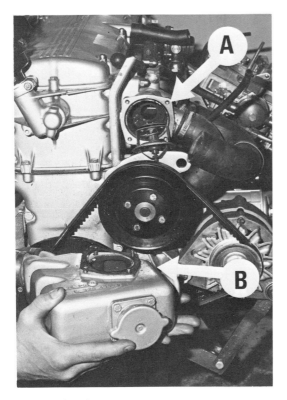

Fig 18. The Thermostat Housing is usually found at the top front end of the engine (A). The thermostat itself is resting on the fresh water pump housing. On this BMW unit, the header tank (B) has been removed to get at the thermostat. The index finger rests on the header tank pressure cap.

with the design of the raw water cooled engine is that, although the engine should warm up quickly in order to run efficiently and smoothly, some water must bypass the thermostat to cool the exhaust system. Water cooled exhaust systems are discussed in detail in Chapter 5. Designers try to keep the temperature differential between the cold incoming raw water and that near, say, the exhaust valve, as reasonable as possible. Incoming water is gradually warmed by passing through, perhaps, a gear box, oil cooler or manifold, before entering the really hot part of the engine around the combustion spaces. If this were not done, the severe local cooling on the hot parts of the block would set up stresses enough to crack the block. Mercury have an extra centrifugal pump on a number of their raw water cooled engines and this does give a more even flow distribution and, therefore, a more even flow temperature distribution round the engine block. I have found that engines with a single pump and thermostat tend to run on the cool side and, overall, do not have the life of a fresh water cooled unit. It is as detrimental to an engine to run it too cool as too hot – and it wastes a lot of petrol.

Thermostat Location and Function. (Fig 18). Most marine petrol engines have the thermostat housing well up towards the front end of the engine. Sometimes the thermostat will actually be in the header tank in an indirect system, controlling fresh water circulation. More often, it is found on the main block just

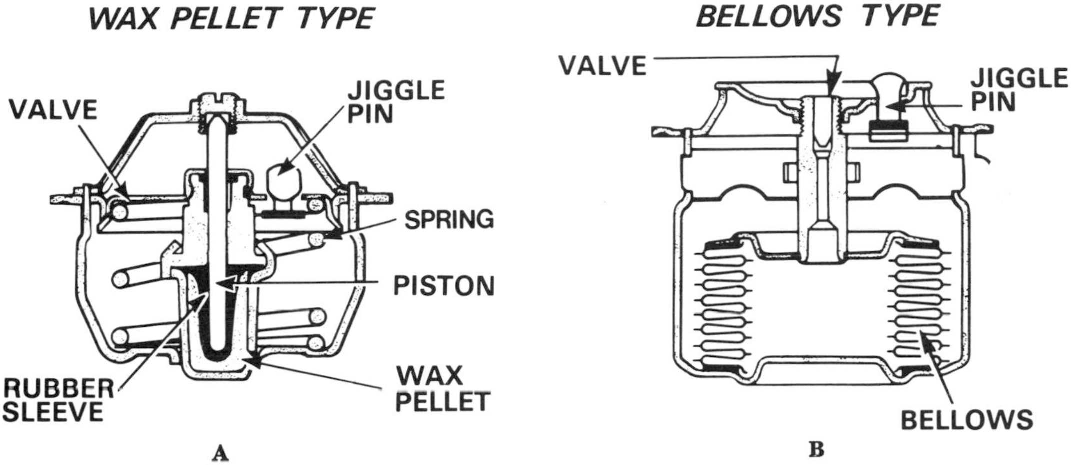

WAX PELLET TYPE

VALVE

JIGGLE PIN

SPRING

PISTON

RUBBER SLEEVE

WAX PELLET

A

BELLOWS TYPE

VALVE

JIGGLE PIN

BELLOWS

B

Fig 19 A and B. Thermostats–wax pellet and bellows types.

in front of the rocker cover or in a similar position in its own housing. This usually makes it easy to get at for servicing. Nearly all engines these days have the wax pellet type thermostat as shown in Fig 19a. In this, a valve remains closed to restrict cooling water flow until the coolant reaches a pre-determined temperature, at which time a small quantity of wax contained in a heavy gauge copper cup expands rapidly. This exerts pressure through a synthetic rubber surround onto a specially shaped piston which is fixed at its top end to a rigid frame. The wax pellet is then forced downward, opening the thermostat valve which is at-tached to it, against spring pressure. The thermostat on the direct cooled engine then allows the coolant (that has been contained within the engine block and manifold) out to the water injection bend for exhaust cooling, while more raw water – prewarmed but still lower in temperature than that ejected – comes in to augment the water in the main block. In the indirect cooled engine, the closed thermostat restricts coolant flow in the engine block until operating temperature is reached, and then allows full circulation to the heat exchanger. When the temperature gauge indicates either abnormally high or low coolant temperatures and the system is generally in order, the thermostat must be suspect. The usual problems are:

(a) The wax pellet fails leaving the valve permanently closed.

(b) Trapped dirt or rust sludge may prevent the valve from seating properly, so causing the engine to run on the cold side.

(c) Pollutants may be deposited on the moving part of the mechanism causing it to stick or operate sluggishly. This results in inordinately lengthy warm up or shows as slight overheating, depending where the valve is sticking or at its most sluggish.

Older engines used the bellows type thermostat shown in Fig 19B. The metal bellows expanded or contracted to move the piston that opened or closed the thermostat. The main problem was that the bellows tended to corrode and once they started to leak they failed to work, and needed replacing. The modern wax type is simpler and more reliable.

Testing and Replacing Thermostats. You can be badly burned by scalding water from the cooling system if you unthinkingly remove a hose from the raw water system, or a pressure cap or hose on the fresh water system when the engine is hot. Take care, then, when setting out to service the suspected thermostat, and always use a thick cloth over the pressure cap if you have to remove it when the engine is hot. Drain the water down and remove the thermostat housing. When removing the thermostat do not damage the thin metal seating which is often sealed with either a commercial sealing compound or a gasket. A thin knife blade run round the edge will usually loosen it. Do not exert pressure by levering it off roughly as this can distort and damage it. When it is removed, check if the valve is seating properly before you do anything else. Hold it up to the light and see if there is a gap between the valve and its seating. If the seal is perfect except for one or two points, it may be acceptable. Next, clean up the unit with a small toothbrush or a moderate air blast from the air line at a garage. There are sophisticated thermostat testing devices with built-in thermometer, but a small pan of warm tap water, a pair of pliers and the old thermostat will be quite adequate. Starting with the water below operating temperature (found stamped on most thermostats) warm the water on a stove. Holding the thermostat in the pliers, immerse it in the water. As the temperature increases the thermostat should begin to open. If you have a thermometer, read off the temperature when the valve begins to open. Don't let the thermometer touch the bottom of the pan but get a true water temperature near the thermostat.

The reading should be within 2°C or 5°F of the marked temperature, and the valve should continue to open progressively until it has travelled about $\frac{1}{4}$in to its fully open position within a further 11-12°C or 20-25°F. Allow the thermostat to cool down and note how it closes and if it is fully closed at a temperature 6°C or 10°F below its stamped specification. If the thermostat fails this simple test, replace it. After cleaning off

Fig 20. This impeller pump is delightfully situated for easy servicing at the top of the engine block and driven from the overhead camshaft of the BMW power unit. Look at that core plug though, situated immediately above the flywheel where it will be hidden in the bell housing. It's a good job that it's stainless steel.

gasket surfaces on the cover of the thermostat and its housing, insert the new one with the same type of gasket as before. This may be a gasket supplied by the maker or simply a proprietary gasket paste. Be sparing with pastes.

Core Plugs. On the main engine block casting you will find a number of small openings in the casting covered with a circular disc or core plug. These are nowadays made of

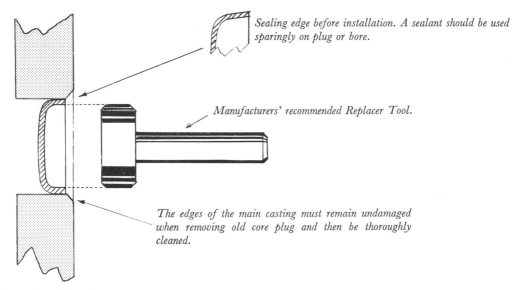

Sealing edge before installation. A sealant should be used sparingly on plug or bore.

Manufacturers' recommended Replacer Tool.

The edges of the main casting must remain undamaged when removing old core plug and then be thoroughly cleaned.

Fig 21. Core plug replacement.

stainless steel and should give no trouble, but older engines may have ordinary mild steel core plugs which in time weep and corrode away. Raw water systems are particularly vulnerable. The plugs are deliberately made to have lower strength than the casting so that, theoretically, they will blow out with excessive pressure, especially that produced by freezing water. In fact, core plugs seem to remain perfectly in place when it is freezing – it is the cylinder block that cracks. Mercury, for example, advise that cylinder blocks suffer horizontal cracking when water freezes in the block, the four and six cylinder models

cracking at the sides of the engine just below the core plugs or along the upper edge just below the cylinder head. Annual inspection of the plugs is well worthwhile, though you will find them hard to get at (Fig 20) if they do need attention. Fig 21 shows a cup type core plug and the methods of fitting. Plugs are best removed by centre drilling and then using a drift punch or pin punch to get them out. Take care not to chip the edge of the main casting hole and see that corrosion is thoroughly removed from the seating in the casting before fitting the new plug. The plug and bore should be lightly coated with a

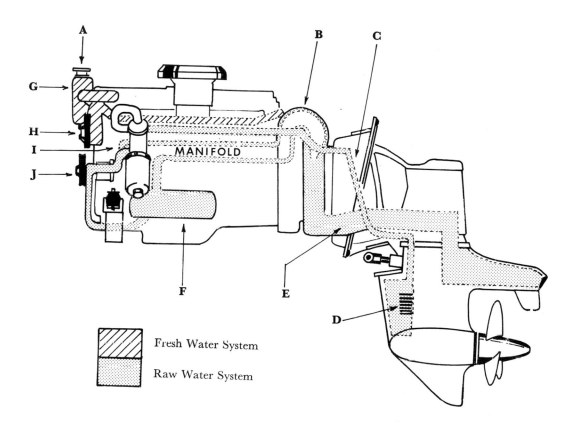

Fresh Water System

Raw Water System

Fig 22. Componenents of an indirect cooling system on an outdrive power unit (based on BMW).
A) pressure cap. B) water injection bend for cooling exhaust gases. C) flexible rubber water intake hose. D) raw water
intake on outdrive; see fig 24. E) exhaust rubber bellows. F) heat exchanger. G) header or expansion tank for fresh water
side of system. H) fresh water pump–usually centrifugal type. I) oil cooler; this is on the fresh water side of system;
sometimes a gear box oil cooler is on the raw water side. J) raw water pump is usually of the impeller type.

gasketting sealant before driving the new plug home. To avoid damage, you must use the correct tool.

Indirect or Fresh Water Cooled Engines

This system is costly to engineer because two separate water systems are used to cool the engine, with a heat exchanger to remove heat from the fresh water side and transfer it by conduction to the raw water side. It also has a header tank to cope with water expansion in the closed fresh water side.

Fig 22 shows a typical flow diagram of the indirect system based on the BMW outdrive power unit. Don't be concerned if your engine is not of this make, as it is purely a matter of identifying the basic parts of the cooling system and knowing their function. Here the raw water intake (D) is on the outdrive leg but on a conventional inboard engine which drives propellers on shafts, you would find a sea-cock near the engine. Whether direct or indirect cooling is used, if the engine temperature soars suspect blockage in the raw water inlet. Plastic ropes, line and bags are a menace, although mud may be the culprit if you have gone aground.

Fig 23 shows a sea-cock which can be turned off while the water strainer above it is cleaned. The gland (D) may eventually weep and need repacking with material from a plumber's merchants. Old packing can be

removed with a hooked wire. Inset is a flushing device that can be used on any raw water side at laying-up time. It fits directly on top of the valve in place of the normal cap. Have it made with several different diameter hose lands so that various water hoses can be clamped onto it. This device allows the raw water water side of the engine to be flushed while the boat is still afloat. Mercury make a special hose attachment which clamps onto the outdrive leg (Part C–73971A1), but with a little ingenuity you can make up a device which fulfils the same function, such as shown in Fig 24. This would enable you to flush the BMW power units from the outdrive water inlet. Note that for all outdrives with this kind of intake, you have to have the boat ashore before you can carry out the operation. ALWAYS REMOVE THE PROPELLER BEFORE STARTING THE ENGINE FOR FLUSHING. A flailing propeller could be lethal. Some separate raw water filters have see-through tops so you can check them each time you turn the main sea-cock on. The filter mesh should be removed for cleaning from time to time.

The metal used for sea-cocks or other types of water intake is important for, if corrosion causes failure, the boat will be flooded and sink. Top quality skin fittings are gunmetal castings (to BS 1400LG2 in UK), but even the best are subject to electrolytic or pitting corrosive attack. I have even come across biological reaction, where heavily polluted water ate deep pits into an underwater valve

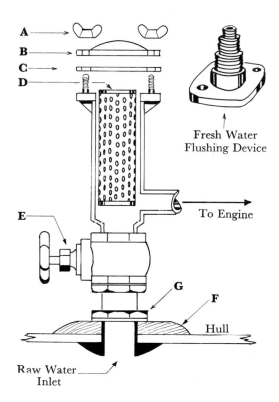

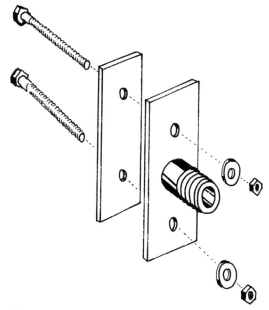

Fresh Water
Flushing Device

To Engine

E

G

F

Hull

Raw Water
Inlet

Fig 23. Raw water inlet and strainer.
A) wing nuts. B) access cap. C) gasket. D) strainer–clean
regularly. E) gland–check for weeping–repack when
necessary. F) backing pad–provides reinforcing and land
on all types of construction; check integrity of pad and
nut (G). G) securing nut.
Inset: fresh water connection flushing device to take place
of (B).

Fig 24. Flushing Device for raw water section of
BMW outdrive unit.

seating. I consider skin fittings to be the most
vulnerable part of the hull, especially when
considering the watertight integrity of the
modern fibreglass or ferro-cement hull. When
my boat is on moorings, I keep the sea-cocks
closed. When laid up ashore they can be left
open to ventilate the bilges. Sea-cocks and
other water intakes are generally connected
to the engine with plastic or rubber hoses
and later I will detail their special servicing
requirements.

The Heat Exchanger is basically a casing containing the fresh water around small bore tubes. The raw water is forced into the casing from the impeller pump and, having flowed through the tubes and taken heat from them, is ejected overboard via the exhaust system. In a dry exhaust system the raw water is simply ejected overboard. The centrifugal pump circulates the fresh water through the casing in the heat exchanger, and so the heat transfer takes place. The only problem likely to be encountered with the heat exchanger is clogging with debris such as mud, corrosion particles or disintegrated lumps of impeller. I use a plastic knitting needle to clean out the tubes, taking care not to damage them. Always work in the opposite direction to the normal flow when removing debris.

Generally speaking the heat exchanger only needs servicing every two to three years, dependent on how much contamination there is in the raw water side and the way you look after the fresh water side. In waters where there is a lot of lime or when the fresh water side is allowed to become contaminated with corrosion or other debris, the efficiency of the heat transfer becomes seriously depleted. On such occasions, if a good scrub and the knitting needle are not effective, a bath of caustic soda solution – preferably boiling – will loosen foreign matter adhering to the tubes. Do not overdo this and wash the units thoroughly before reassembly. You will need sets of gaskets to rebuild the heat

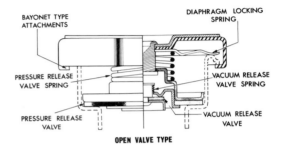

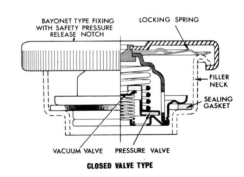

Fig 25. Header tank pressure relief caps; open and closed valve types.

exchanger.

Because heat exchangers are made of different metals in the tube stack and the outer casing, some manufacturers (Mercury for example) provide cathodic corrosion protection. This is usually a small zinc block electrode inserted into the cooling header tank or, on some of the Mercury units, in the body of the heat exchanger. Be certain you renew this protection.

The Header Tank and Pressure Cap. Checking the level of the fresh water in the header tank should be a routine task before you start the engine. When you remove the pressure cap on the coolant tank, give it the respect it deserves. It is a scientific bit of engineering designed to do a specific job besides keeping the coolant in the tank. By pressurising the fresh water side of the system, the engine is allowed to operate at more efficient higher temperatures. This pressure causes the boiling point of water to be raised. This is why the coolant boils if there is a broken hose or other leak in the system. As the engine temperature rises, evaporation and expansion of the coolant takes place and because the system is sealed by the pressure cap, an internal pressure builds up and affects the whole of the fresh water system. In Fig 25 the main pressure release valve is held against its seat by a spring with a predetermined strength which controls the pressure at which the valve functions. If the cooling system pressure rises above the rated pressure of the cap the relief valve is forced off its seat and the excess pressure is relieved via the overflow pipe. In addition, a small lightly loaded valve operating in the reverse direction to the main valve, prevents any vacuum building up in the system as it cools. Damage to the pressure cap is usually through mishandling or other rough treatment. Some pressure caps have a composition sealing washer, and this can be damaged with oils or some anti-corrosion aerosol sprays. Handle the cap with care and keep it clean with a little fresh water and an old toothbrush.

Whenever there is boiling or excessive loss of coolant, check the pressure cap, remembering the warning about scalding. Always place a thick cloth right over the pressure cap area if you intend removing it when the engine is hot. Some caps have an excellent built-in safety device. At the first part of the turn, the pressure is released while the cap is held firmly in place so that it cannot spew boiling water in your face, and in the second part of the turn it comes fully away. You should still hold a cloth over it for safety. With a good cap you should hear an initial pressure escape, provided there is no leak in another part of the system, such as at a hose connection. If the valve seal or gasket (if employed) looks suspect, fit a replacement as they cost very little. Spares should always be carried. AC Delco make a pressure cap testing device and this may be available at your local garage. Even when the cap appears to be good, there is, for the amateur, no way of knowing if the valve spring is of the correct tension or if it is releasing pressure lower than it should.

Hoses and Clips. The cooling systems of engines could not exist without rubber hoses and clips to connect the various components. Hoses perish–especially those carrying higher temperature cooling water. They harden with age and crack. The hotter they get the quicker they go. The engine designer may have had a bit of forethought and designed hoses of the same diameter for the cooling system, in which case it is possible to have in stock the main specialised bends and one straight length of hose that can be cut into any length. The more awkward engines mean you have to carry a variety of angles and bores. Pushing the hose onto the spigot would appear to be an easy job but it is amazing that even such a simple operation can be bodged. Ensure that the axis of the hose is in line with that of the spigot. Lubrication with a washing up liquid will help ease the hose on. Wherever possible use stainless steel clips, two being safer than one.

If you do not carry a reasonable supply of shaped hose and straights, the least you can do from the point of view of safety is to have some means of making a temporary repair. A small tube of one of the rubber sealants or even chewing gum can be forced into and across a split, and then bound with electrical insulating tape; this should suffice to get you home. Any rubber compound hose which comes into contact with mineral oils will show accelerated deterioration. Avoid oil spillage onto hoses and, if you use emulsifying oil for inhibiting the cooling system, check hoses for aging.

As hoses age they shrink in diameter, and it is normal maintenance practice to go over them all a couple of times each season to see that they are tight. Don't overtighten them so that the clips cut into the hose. I double clip all hoses from the sea-cock to the raw water pump. In fact any pipe below the waterline should be double clipped. Watch out for salt weeping round the joints of raw seawater pipes as this can be a sign that the pipe under the hose is corroding. Copper pipes show green verdigris salt crystals – a condition that must be taken seriously.

Plastic hoses. Although most hoses on the engine are rubber reinforced with canvas plys, plastic hoses are used by boatbuilders to connect the raw water supply from the sea-cock to the raw water pump. This is all very well if the hose is well clear of a hot spot such as an exhaust pipe. Heat has been known to melt a badly installed pipe and cause a new boat to sink. Check pipes such as these and see that they are secured in such a way that the engine can flex without pulling them off, but that they cannot move accidentally onto a hot area. Plastic pipes age-harden but there is also, in time, chemical hardening in the presence of fuel and oils.

Winterizing Cooling Systems

Preservation of engines during long periods of non-use should be considered as vital if they are to be reliable and long-lasting. Procedures are as follows:

Raw Water Systems. While the boat is afloat, run up the engine and, preferably, go for the last spin of the season. As soon as you are re-berthed, or the engine has run for a few minutes at operating temperature, close the sea-cock and fit a connection plate (already mentioned) over the top so that fresh water can flow to the engine. The engine flushing water can come initially from a hosepipe ashore if there is a tap or a five gallon drum. If you use a drum, the time you run the engine must be limited so that you do not run it dry. If using a shore supply, you can flush the system out thoroughly, so removing a lot of salt as well as other pollutants from a sea-going boat. You haven't finished yet, though. Fill up the drum again, but this time add an emulsifying oil such as Duckhams Aquicut 40 or Shell Donax C., Esso Cutwell or similar: oil to water should be about 20 per cent. When the engine was at its correct operating temperature after the last run, you should have drained off all engine and gearbox oils and replaced them. I will say more about oils for winterization in the chapter on lubrication. However, you now run the engine again with the new oil in and connected to the cooling water drum until the solution is all used up; then stop it. While the engine is still hot, disconnect the drum, replace the top on the sea-cock and open all drain plugs on the engine. The idea is that a thin film of oil will dry on the internal metal surfaces to preserve them. Note that such oils are NOT anti-freeze items and must be thoroughly drained down. Brass drain cocks are used almost universally on marine engines for draining cooling water. Their location is often difficult and you must familiarise yourself with every single one. They have a nasty way of being behind things and are difficult to turn even when they are in good condition. I treat them by spraying with LPS before I open them so that the metal surfaces nearby are protected from the water. I also have a cocktail stick or a bit of wire handy to poke them out immediately they are opened, as a lot of debris tends to gather behind them and prevent water flowing out. If the debris is really bad, the drain cock should be removed completely until water flows out of the threaded hole cleanly. You can then service the drain cock. If you are draining down a boat already laid up ashore, make quite certain that she is plumb, and not bow down on the cradle or props. If she is, there is danger that water draining down from the engine will not all come out. The result will be frost damage – cracked cylinder blocks and exhaust manifolds.

Indirect Systems. Both sides of the system, raw and fresh, must receive treatment before laying-up. Engines with outdrive units need special care as both the engine and the water intake and passages on the outdrive need attention. First, the straightforward inboard engine. Like the raw water unit it is sensible to see that the raw water side is flushed out as already described. Personally, I do not like running engines when the boat is ashore, since there is already enough stress on the hull without adding that of a vibrating engine. With the majority of outdrive units, there is no option but to flush the raw water side when the boat is ashore. Generally, you have the option of leaving the fresh water section dry or wet. By dry, I mean that, like the raw water section, the old coolant can be drained down and the same emulsifying oil mix put into the system for a final run, before draining down and leaving the engine completely dry. Leaving the system wet means leaving a correct mixture of water and anti-freeze in the system so that it is protected from the worst winter frosts. Mercury insist that their 470 engine has to be winterized this way, but you should check the manufacturer's recommendation and conform to it. There are good arguments, in fact, for leaving a system wet, with anti-freeze protection: for instance gaskets do not dry out, surfaces are protected with the rust inhibitors in good quality anti-freeze products, and the engine is that much easier to put into service next season. In engines where there is a lot of light alloy, the corrosion inhibitors do a really good job and I would recommend laying-up wet. Items like light alloy water pumps respond to decent treatment. The disadvantage is that anti-freeze costs money and it has a habit of seeking out any weakness in a hose or joint. However, this may indicate that one of the engine hoses is at the end of its life anyway. Over a number of years I have become convinced that leaving the fresh water cooling system wet, with the right degree of anti-freeze protection, is to be recommended. A good quality anti-freeze, such as Smiths Bluecol, consists of pure ethylene glycol with added corrosion inhibitors. A top grade inhibitor only needs renewing every two years, provided the cooling system is in good mechanical order and clean, and that proper procedures are followed to keep the specific gravity of the anti-freeze solution at the correct level. With Smiths Bluecol, a specially graduated hydrometer (Fig 26) is available, complete with a slide rule so that you can determine the degree of frost protection required and, from hydrometer readings used in conjunction with the slide rule, the quantity that must be added to maintain protection. The anti-freeze solution only becomes diluted when the header tank is topped up with water during the summer.

Fitting Out Cooling Systems

With both raw and indirect cooling systems, start at the seacock and ensure the top is assembled with the correct gasket under it and that the cock is shut off. After checking the sea water hose from this cock to the raw water pump, reassemble the impeller pump. Replace the impeller, then the end plate gasket and end plate. I like to work a little grease sparingly into the paper gasket to give a good seal and prevent corrosion between the body of the pump and the end plate. Check all hose clips and rubber pipes. When you are sure that the whole system is together on the raw water side, you should be ready to go. On indirect systems where you have left the

Fig 26. Checking the specific gravity of anti-freeze with a Smiths Industries 'Bluecol' tester.

lock is prevented. An air lock will cause over-heating. Loosing off a plug slightly you may hear the trapped air hissing out. As soon as the water appears, close the plug down securely. I have been in a panic more than once when, after the engine has been started for a few minutes, the temperature gauge has been too high because I did not take enough care to get the air out. However, at first starting, the wise owner watches each part of the system in turn, checking that there are no leaks and everything is functioning correctly. In indirect systems left wet, you need only check water levels and top up the header tank after recommissioning the raw water side.

fresh water side drained down, you will have to see to that part as well. The fresh water side needs filling up. Check your engine handbook, as quite often it is necessary to remove drain plugs from cylinder blocks or plugs from a header tank to ensure that an air

four

Lubrication

The job of oil is to keep apart metal surfaces which are moving relative to each other in the engine. No matter if one surface is static and one moving, or both are moving, they must be kept separate by a minute film of oil. Oil must also inhibit corrosion and protect the metal surfaces from chemical attack. The acid products of combustion can produce alarming effects and, if it were not for the oil absorbing these corrosive products and carrying them away to filters and sump, they would have a hearty meal of the engine's inside. Oil is the lifeblood of the engine, and you must follow the engine maker's oil specifications as he will have done exacting tests to ensure the unit has a good service life.

Lubrication of Two-Stroke Engines

In Chapter 1, I mentioned briefly the main difference between two- and four-stroke engines. Here I must go into some aspects of lubrication in a little more depth as they can make a great difference between a sweet-running, trouble-free engine and one that would make the Ancient Mariner's albatross gloat. Generally, petrol two-stroke engines are found at the low horsepower end of the market.

Until recently, when the oil companies developed special two-stroke oils, ordinary lubricants had a bad name for sooting up and closing the gap on spark plugs. Problems arose because oil was put into the tank first and petrol added slowly afterwards so that it never mixed thoroughly. A fast pour, however, usually did the trick. Another fault was that petrol/oil mixtures were left in tanks for long periods, especially over the winter. In such cases, the specially added aromatics merely evaporated and left less than perfect starting properties. Always remove fuel from tanks and store it in containers with a little air space left at the top. Even today it is essential to add the correct amount of oil if you are to avoid problems. It's no use thinking that being generous with the oil will make the bearings last longer.

Some of these old engines had chains to link starting handle and flywheel. Any external chain needs regular cleaning. Remove it from the engine by undoing the spring link and then agitate it in a bath of paraffin. Let it drain down then re-lubricate with a chain or engine oil of about SAE 30 before replacing.

The rest of the lubrication points on a two-stroke are small ones, such as on magnetos, gear handles, gearbox and the like, and these are mentioned in appropriate sections of the book. Winterize the cylinder and crankcase by adding a tablespoon of oil to each cylinder and distribute it by turning the engine over.

Lubrication of Four-Stroke Engines

This is much more complex than the previous section, involving oil pumps, filters, pressure

−23°C	−12°C	0°C		16°C	32°C	Typical applications
		10W/30				Volvo Penta BB1000, MB 18, BB30, MB10A Watermota Super Shrimp
		10W/40				Volvo Penta 170/270, 130/270, 115/100, 140/280, 240/280, 190/280, 120/270
		10W/50				
		20W/40				Volvo Penta BB115A, BB170A
		20W/50				BMW B190/220
		30				Watermota Sea Wolf, Tiger, Leopard Volvo Penta C5
10W						
		20W/20		40		
−10°F	+10°F	+32°F		+60°F	+90°F	

Table 3. Guide to Oil Specifications for specific ambient temperatures
Note: the old MS specification is now replaced by SE.A11 SAE

relief valves and a variety of oils. Table 3 gives a basic guide to oil types based on ambient temperature. Ambient temperature, in our case, is that of the engine compartment, which under normal running conditions is likely to be higher than the air temperature outside the boat. Always, however, use the engine maker's specifications, and stick to the same brand of oil, as different makes are not always compatible. Fig 27 shows the basic

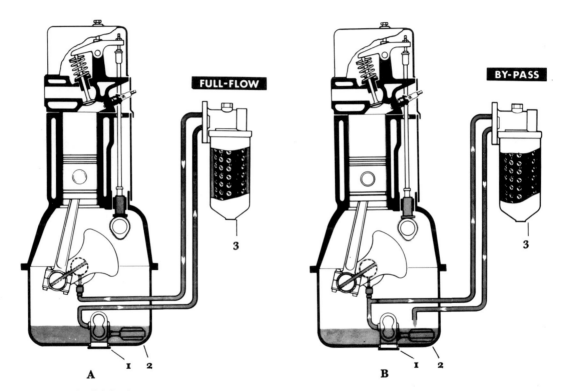

Fig 27. Engine lubrication systems.
A) the almost universally used modern 'full flow' system. B) the older 'partial' flow system.
1) Oil pump; either a gear pump or more commonly an eccentric rotor pump type is used; usually only need overhaul at major overhaul times. 2) Coarse filter for large particles only. 3) Main filter; may be of renewable element type or the throwaway canister type.

elements in the lubrication system.

The full flow system is now the most universally used system for filtering engine oil. All the oil passes through the filter from the pump where it is cleaned and, still under pressure, directed to all the main engine bearings and the valve gear in the cylinder head. Older systems used a bypass system which only allowed the filter to clean a proportion of the oil in each circulation. Usually this was about 10 per cent per hour. Its decline has been caused by the great improvement in both filtering media and the quality of modern lubricants.

Oil pumps need not concern you too much as they are usually well out of harm's way tucked up in the engine sump. The two types of pump for lubrication duties are the gear pump and the rotor pump. They generally need no attention except during a major overhaul. Oil from the pump is fed by means of pipes and drilled passageways to the engine's main bearings, to internal drive chains and the camshaft. The oil is filtered to remove debris and chemicals which would otherwise soon cause blockage. The unseen pollutants include metal particles ground off the surfaces of bearings, varnish, salt, dust, silica, carbon, acids – a collection which would soon destroy the engine if not removed. When the engine is running slowly, fuel is scraped down by the pistons to dilute the oil. A quick burst of speed to get the engine really hot will evaporate this fuel, but even then it can lead you to believe that the engine has suddenly begun to consume oil when you check the dipstick. Another 'drag down' product is water. For every gallon of fuel burnt, the engine produces about a gallon of water. Most of this is flung out through the exhaust but some does get past the pistons and into the sump. Leading oil companies say that up to ninety per cent of engine wear

Fig 28. The work of an oil filter element.

comes from this water turning acidic in combination with the products of combustion, and attacking the metal inside the engine.

Fig 28 shows what an AC filter does. This should be sufficient to convince you that all dirty oils should be drained down from the engine gear box and reduction gear or outdrive at the end of the season. To leave them in the engine is to allow the sludge to drop out and coat internal surfaces, acids to eat their way into cylinder walls, and corrosion to take place at a much higher rate than with clean oil in the engine.

There are two basic types of filter in common use. The first type has a filter element as shown in Fig 29, where the outer metal casing has a central spindle onto which the renewable element is placed. This is the BMW

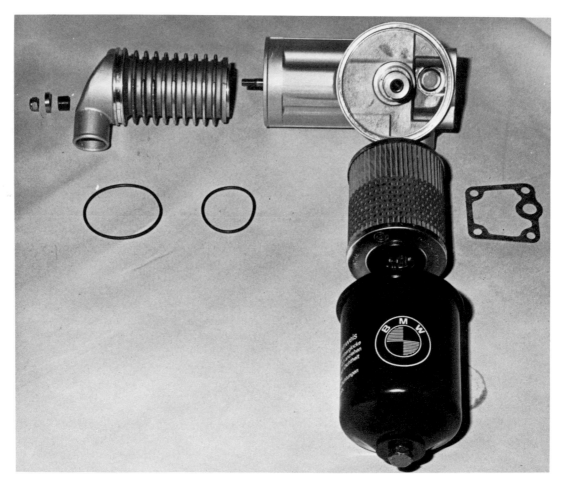

Fig 29. The oil filter and oil cooler of the BMW (6 cyl) engine.

unit which has a cooler incorporated above it to cool the cleaned oil returning to the engine. Many engines have a separate oil cooler working on the same heat exchange principle. With element oil filters it is important to renew the gasket which is usually supplied with a new element. The gasket is generally of square section synthetic rubber and fits into a groove in the top casting. A pin or knife blade is the best instrument for prising the old one out. The tricky thing about replacing the gasket is pushing it home into the groove without getting a twist in it. If you are a bit careless you will get an oil leak. I prefer to replace the element at the end of the season when new oil is put into the engine.

Changing any type of oil filter is a dirty business and Fig 30 shows the least messy way. A big plastic bag is placed under the filter to catch the drips of oil and the old filter element. Small plastic bags will catch oil from the gear box or reduction gear as with most marine engines the position of drain holes makes it impossible to fit a tray or other receptacle underneath.

Some filters, such as the AC Full Flow type illustrated in Fig 31, incorporate a ball relief valve which occasionally might give trouble. Normally, as you will see from the drawing, the cleaned oil flow bypasses this relief valve on its return to the engine. If, however, the filter is becoming blocked, perhaps through being over-impregnated with dirt, the oil will bypass the filter element and proceed to flow via the relief valve. It is safer to have a little

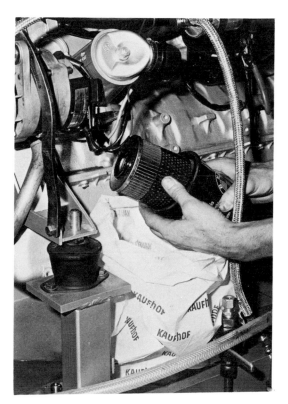

Fig 30. Changing an oil filter element; have a plastic bag under the oil filter to catch the dirty oil when you are changing it.

dirty oil for lubrication than none at all.

The second type of filter is the throw-away canister type. In these, the whole canister is unscrewed from the filter base on the engine and disposed of. Although they are not sup-

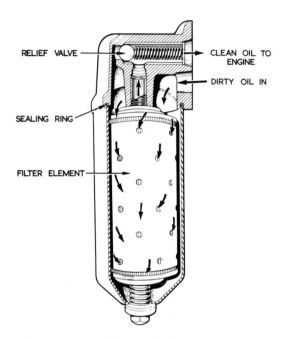

RELIEF VALVE

CLEAN OIL TO ENGINE

DIRTY OIL IN

SEALING RING

FILTER ELEMENT

Fig 31. The relief valve in the AC-Delco full flow oil filter.

posed to spill oil when they are undone. you really need to have a plastic bag underneath to catch drips and the old canister. The sealing gasket on the bottom of the canister is fixed and only needs a smear of grease on it to prevent it sticking. WIPAC and several other firms make a neat tool to grip these filters and remove them easily. The tool slips over the top and a toggle allows a good grip to be applied.

Theoretically, you should only need to hand-tighten the canister so that it comes off easily at the next change. The tool is a worthwhile addition to the tool chest.

The frequency for filter changes depends on how many engine hours you do each season. If the sailing man does a hundred engine hours, that is about average and one filter change at the end of the season will suffice. The power owner may do a good deal more and so have to perform changes in mid-season. An engine hour meter is a useful accessory if you do not keep a log.

If you do not use a filter that is standard equipment, make sure you get a precise equivalent; it might appear the same but could have a coarser or finer filtering medium. This means that with the finer medium you would have to change the filter every fifty instead of one hundred hours; a costly business but acceptable if no alternative is available. A coarser filter should only be used in the most dire circumstances – and then for the shortest possible time. Fig 32 offers some tips on draining oil from an engine.

Specialist Oils

There are a number of useful specialist oils. They are an additional cost but, in given circumstances, are indispensable.
Flushing oil is a light bodied oil often fortified with solvents capable of removing carbon,

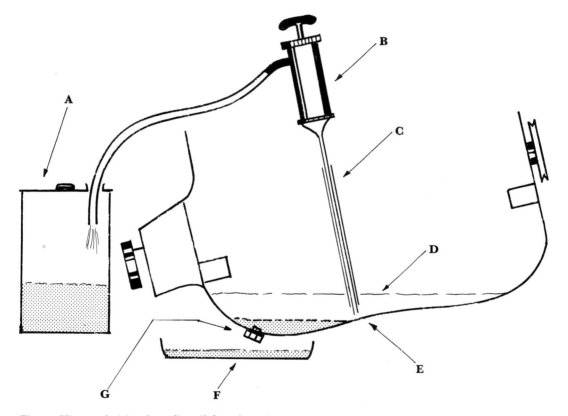

Fig 32. Hints on draining down dirty oil from the engine.
A) Always carry aboard an empty old oil tin to drain into. B) Special sump pumps are made, often with narrow bore intake pipes that will fit down the dip stick hole. C) Dip stick hole. Gear boxes, reduction gears and outdrives may be emptied via the oil filler holes if they cannot be directly drained via a plug. D) Normal oil level will vary fractionally with installation angle of engine in your particular boat. E) From (D) note that all engine oil would not be drained and would leave the worst heavy particles in the sump. F) Old oil cans can be cut to make shallow sump trays for getting the last oil out; often this tray would spill if you tried to get it out so use that oil pump again; sheet polythene can be used if supported to catch the oil instead of a tray. G) Sump plug. Never overtighten and check the washer if it has one.

sludge and varnishes from the interior of the four-stroke engine. It is particularly useful when taking over a secondhand boat if you don't know how well it has been looked after. Giving the engine a clean fresh start will add to your confidence as well as contributing to the engine's life. With engines you have known from new, a four-yearly flush should be sufficient.

Procedure is quite straightforward. After you have drained off the engine oil at the end of the season, add the flushing oil up to the lower dip-stick mark. Re-run the engine at a fast idle for fifteen to twenty minutes. Don't run the engine under load as flushing oil has only sufficient lubricating properties to work at low speed without stress. Switch off and immediately drain down the now filthy oil and change the oil filter. Remove the rocker box cover and, with a clean lint-free cloth dipped in flushing oil, clean off the tappets and the rocker box cover. Then, with an oil can, flush down the tappets etc. to leave them bright and clean. Refill with either normal grade oil or storage oil.

Storage oil is used in engines which are going to be laid up for six months or over. In the less extreme climates of Europe and the USA it is quite acceptable to lay up the engine with the ordinary grade oil left in the sump. In more extreme climates, and if the engine is to be left longer, it is recommended that a storage oil is used. Castrol Storage Oil, Shell Ensis Engine Oil 30 or Duckhams Adfilm 730 are typical products which work by a combination of contact and vapour protection. These oils should be added up to the level of the upper mark on the dipstick, either after normal dirty oil has been pumped out at the end of the season, or after the flushing oil has been removed. I would prefer to use a flushing oil before long storage to ensure maximum cleanliness within. Run the engine to allow the storage oil to circulate. Final treatment consists of removing the rocker cover and pouring a cup or so of oil over the tappets and introducing about a table-spoonful of oil into each cylinder. Two-stroke engines benefit from a dose of preservative oil in the cylinder, introduced by way of the spark plug hole. Turn the engine over a few times to ensure that the oil has coated cylinder walls and that some is left on the piston rings. Spark plug holes can be shut off for the winter with old plugs to keep the damp air out. Some boat owners use special plugs filled with desiccant silica gel, which has blue crystals that turn pink when they are saturated. They must be re-heated when in this pink state or all they will do is keep the air in the cylinder space damp.

When checking the engine at the beginning of the next season, free the cylinders of oil by turning over the engine on the starter without the plugs in. If this precaution is not taken you could bend the connecting rods by creating a hydraulic lock in a combustion space full of oil instead of air. Preservative oils must be drained off at the start of the season as they are not designed for use as a running oil. Warm up the engine with the oil still in under light running conditions and, after it has been drained down, fill up with the normal grade of oil.

Specialised greases are available for a multitude of applications, but there may come a time when you wish to simplify matters to avoid confusion and expense. I find a top quality lithium based grease, such as Duckhams Keenol, is adequate for most jobs. Few engines these days have grease points, but it is useful when assembling nuts on studs or on parts of the outdrive where grease is specified. Grease must never be used where oil is used and vice versa. Mercury do a really comprehensive range of lubricants for their engines and it is worth knowing about them. Incidentally, I think buying additives to add to premium grade engine oils, as recommended by the engine maker, is a waste of money.

Aerosol lubricants. In recent years a whole army of these has become available and many do a first class job in inhibiting corrosion, cleaning and de-watering electrical apparatus and lubricating. Take care, though, when selecting a spray for some of them harm plastics and rubber – and there is plenty in the electrical sections of the engine.

Over the last five years, I have found LPS products to give excellent service. Their No 1 is a light greaseless metal protector, ideal for throttle cables, electrical systems where it waterproofs as well as de-waters. The No 2 is useful for introducing to the air intake during the last run before laying up, to inhibit the carburettor, intake manifold and valves. The air cleaner must be removed and during the last few minutes of the run, the petrol tap shut off. Gradually the petrol will mingle with the LPS, become a heavier mixture and the engine will stop. An extra squirt or two will finish off the job. Stuff up the air intake with a cloth soaked in LPS 2, sealing it for the winter. LPS 3 is a heavy duty film, ideal for a final spray over the whole engine to protect it throughout the year. No matter what lubricants and greases I used to put on water pump and plates I always ended up with verdigris round the edge, but since I started using No 3 after replacing the impeller at the beginning of each season, the pump has remained like new. Where there are extreme demands on an anti-seize material such as the propeller retaining nuts, on outdrives which may be scoured by sand, mud and saltwater, Loctite Heavy Duty Anti-Seize is recommended. There are other similar products marketed, but be certain they are safe on all parts of the engine before you buy them.

five

Fuel, Carburettor and Exhaust Systems

Technically speaking, this is the most complex subject to write about. The link between each stage of the petrol engine's fuel, from the way it is made to a specific octane rating, its mixing in the carburettor, its burning in the cylinder and its final expulsion from the exhaust pipe involves many variables. You should have a little understanding of these to get the most out of the power units.

Fuel and Octane Rating

The engine designer will have determined the cylinder head shape and volume which will best burn a specific octane rated fuel. The higher he can make the compression ratio, the more he can expect the fuel to burn at maximum efficiency to give the highest power. The compression ratio is that between the total volume of the cylinder and the volume of the combustion space when the piston is at top dead centre. Although it might seem that to gain more power all the engineer has to do is to put up the compression ratio, it is in fact limited by the onset of detonation, 'knocking', or 'pinking' in the combustion of the fuel.

When the petrol/air mixture burns in the cylinder, the shape of the combustion space, the moment of ignition, the temperature of the charge and its pressure, the effectiveness of the cooling system and the mixture strength determined by the setting of the carburettor are just some of the items that influence the onset of knocking. The ideal situation in the cylinder is for the minimum amount of fuel to be mixed with the correct amount of air, so that every drop of that fuel will burn to give a steady pressure rise within the cylinder, thus developing power in the most efficient way. When any of the factors given below interfere with this, the flame starts burning in the cylinder and suddenly produces auto-ignition of part of the charge.

The effect of engine conditions on knock is broadly explained as follows:
Compression ratio increases knock by increasing charge pressure and temperature. I am not concerned with up-rating engine power for racing purposes in this book and, as long as gaskets of the correct thickness are fitted and cylinder heads are not shaved too much during reconditioning, the compression ratio should remain as it was originally intended.
Spark advance increases knock by increasing the maximum charge temperature and pressure. Be certain that the timing is correct to the manufacturer's specification.
Throttle opening increases knock by increasing the charge pressure. This is not usually a problem, unless you are trying to fit a carburettor other than the one specified by the engine maker.
Mixture temperature is important, as the temperature in the engine space will affect it. Elevated temperatures increase knock by increasing charge temperature. Bad design, where hot exhausts run near fuel lines, fuel tanks becoming affected by sunshine, high

ambient temperatures – these are just some of the factors within your control. You can at least ensure that fuel gets to the engine at a reasonable temperature. The Coast Guard Rules for Tank and Fuel Installations are mandatory in the US but would make sound practice anywhere. Certainly, the installation of spark and gas proof extractor fans and the provision of a plentiful supply of fresh cold air to engine spaces should have high priority – much higher than it usually receives. The problem is that fans are out of sight and not as appealing as the bits and bobs of nauticalia plastered on cabin bulkheads.

Engine speed reduces knock by reducing the reaction time, but a reverse effect may be caused by incidental changes in other conditions.

Carbon deposits increase knock by increasing surface and, therefore, charge temperatures; also by the indirect effect of nominally increasing charge temperatures. Occasionally there will be a carbon build up that acts as a glow plug, igniting the mixture in the cylinder after the engine has been turned off. This is often referred to as 'dieselling' and indicates that a de-coke is needed. This self-ignition is not to be confused with a much more serious defect – a blown gasket, which will sometimes manifest itself with a dieselling-like run on when the ignition is switched off. The blown gasket leaks hot exhaust from one cylinder to ignite the incoming charge of the adjacent cylinder. Running on demands immediate investigation.

Atmospheric pressure increases knock by increasing charge pressure. Humidity reduces knock by its dilutant effect and, by virtue of the high specific heat of water vapour, reducing charge temperature.

To complicate matters, governments in different parts of the world have, in recent years, reacted to the problem of pollution, especially of lead, in different ways. There is still controversy as to whether lead emission is bad for health, but engine manufacturers have had a headache designing motors to conform to the various types of fuel available. There is a worldwide trend for emission control on the lines laid down in the USA, but at the moment more concern is directed towards fuel conservation measures. I would recommend contacting the manufacturer if you are in any doubt as to the correct grade of fuel to be used in the engine. Mercury Marine are very specific in their recommendations. They say that 'Valve seat failure may occur from using lead-free or no-lead gasolines.' On their engines made before 1974, 'Use any good grade automotive regular leaded or premium gasoline with a minimum average octane rating of 88 (93 Research). In areas where leaded premium and non-leaded regular are the only gasolines available, leaded premium should be used in the initial break-in period. Non-leaded regular gasoline may be used after the initial break-in period is complete, provided a minimum of one tankful of leaded gasoline is used after each four tankfuls of non-leaded. 1975 and newer models can use any good grade automotive regular leaded or non-leaded or premium gasoline with a minimum average octane rating of 88 (93 Research). All models may use 86 average octane (90 Research) leaded gasoline (with 0·5 to 4·0 grams lead per gallon) if the gasoline described above is not available, provided the ignition is retarded 4° to prevent harmful detonation.'

There are points in the above quotation that need clarification. First, the confusing business of octane ratings.

For years there has been a number of ways of determining the octane rating of petrol. Some are based on laboratory methods where the maximum controlled conditions can be imposed, and others on road tests of vehicle engines where many variables mentioned earlier in the chapter have to be contended with to get reasonable results. The octane scale is really a comparison of fuel with two standard primary reference fuels: heptane, which has the low value and is defined as having a zero octane number and iso-octane, which has a high anti-knock value and is given an octane number of 100. The octane number, or knock rating, of a fuel is the percentage by volume of iso-octane in a mixture with normal heptane which gives the same intensity of knock as the sample fuel when tested under the same conditions. The motor method of testing is more severe than the research method, hence the Mercury recommendation of 88 motor being similar to 93 research. Generally, marine engines are built with modest compression ratios that are happy on mid to low octane rated fuels. High performance engines with higher compression ratios will demand a premium grade.

There is much work going on in fuel development and rating, but this simplified explanation will serve to show that a suitable petrol must be used with a marine motor.

If further proof is needed, refer to Fig 33 which shows the results of pre-ignition and detonation or fuel knock. Pre-ignition is abnormal fuel ignition caused by combustion chamber hot spots. Control of the start of ignition of the gas/air mixture in the cylinder under compression is lost. The combustion pressure rises too early, part of its total power being lost against the still rising piston. The result is a loss of power and rough running. Big end and little end bearings suffer excess pressure and the increase in combustion chamber temperature can burn the piston and valves, especially the exhaust ones. The hot spots in the combustion chamber are caused by glowing deposits – the result of using the wrong oils or fuels or a combination of both. They may also be caused by overheated spark plug electrodes when the wrong heat range plug has been used or one is defective. Finally it may be caused by a protuberance in the combustion chamber, such as an overhanging piece of gasket, an improperly seated valve, or any other inadequately cooled section of material which can serve as a heat source.

Fig 33. The destruction of the pistons and engine by persistently running with pre-ignition and detonation conditions.

Detonation can be equally damaging. Sometimes it is referred to as fuel knock or spark knock, or even carbon knock, but it will certainly be detectable to the ear as the fuel burns abnormally with a violent explosion. The explosion will cause overheating sufficient to burn out pistons, spark plugs and valves and, in bad cases, all the phenomena associated with and including pre-ignition. Detonation may also be a result of:

1. Over-advanced ignition timing.
2. Lean fuel mixture at or near full throttle. This may also be due to a leaking carburettor manifold as well as a wrongly adjusted carburettor.
3. Cross firing spark plugs.
4. Build up of carbon deposits in combustion space, resulting in higher compression ratio.
5. Inadequate cooling of the upper part of the engine by deterioration of the cooling system.

Fuel Tanks

The fuel tank that cannot be cleaned is a hazard, and all too often because it is placed in a most inaccessible place; once it is filled it receives little attention. A good installation will have a primary fuel filter to remove debris before it gets anywhere near the carburettor. Even so, the tank will still get a build up of dirt from the filler tanks ashore or from dust ingested through the breather pipe. Although Explosafe reduces explosion hazards, its use makes tank cleaning impossible for the amateur as tank washing procedures, outlined later, are not possible.

You must always take great care when cleaning fuel tanks. It only needs one spark to cause a big bang. Having ensured that the electrical bonding, that would lead any static electricity to earth before it could spark, is in place, you need a means of checking the bottom of the tank for debris. In many countries it is illegal to store petrol or gasoline in other than approved cans and at specific distances from buildings. It is advisable, during tank cleaning, to remove every last drop of fuel and contact your local fire officer if you feel advice is necessary where you have to store anything other than a small quantity. Patay Pumps make a model specifically for fuel pumping duties similar to the diaphragm for bilge pumping. Check with the manufacturer that the rubber parts of such a pump are compatible with mineral oils, or the disintegrating rubber might add to the contamination. There should be a large access plate on the tank top; rid the tank of all fumes before removing it.

Provided you can reach into all parts of the tank, Shell suggest that a solution of Teepol is safe for tank cleansing; the tank is filled with the solution to drive out the dangerous fumes. Then, part emptying the tank will give you sufficient solution to wash the tank round, paying particular attention to the bottom. There will be three main parts to rinse and dry:

1. The fuel filler pipe to the tank; a pull-through rag-mop (clean and lint-free) weighted with a bit of lead will dry the pipe.
2. The tank itself; use a lint-free cloth.
3. The up-take pipe running from the tank bottom.

You can usually blow clean air down the pipe (3) to remove solution from it, if you disconnect it just outside the tank top. Let the whole lot air for a day or two before replacing tank tops and fuel lines.

Particular caution is necessary when cleaning fibreglass tanks. Though the initial Teepol/water solution will have driven out the explosive gases, there is a danger that poor quality moulding will have created internal tank surfaces that become impregnated with sufficient fuel to create an explosive vapour after the tank begins to dry. I have seen fibreglass tanks in such a state of disintegration that the filters were regularly clogged with glass strand debris; if this happens to you then you should get rid of the tank.

One of these days someone will come along with a fibre optic so that the bottom of tanks can be inspected easily. Until then, put this unpleasant cleaning job on a regular schedule.

. Check fuel tank breather pipes, to see that they lead dangerous fumes overboard in a safe place, and that their terminations have not become clogged with dust. Finally, check the gaskets on filler caps as, with decks awash or rain lying on a deck during winterisation ashore, this is one place where water can and does get into fuel tanks. When laying-up it is a good plan to check that the water is not lodging over the filler cap – as it might well do when the boat is ashore and sitting at a different angle than when on the water. A friend of mine who didn't check this was surprised at fitting out time when he took a gallon and a half of water out of his tank.

Even with perfect sealing, condensation can create problems in fuel tanks over the winter lay-up. I don't recommend leaving them filled over the winter, because it can be dangerous and self-defeating for, although the fuel may minimise space for condensation, it deteriorates as volatile parts, especially aromatics, are lost through evaporation.

Fuel Filters and Fuel Lines

Have a receptacle handy when you undo pipes or remove filters for renewal or checking. Fuel must never be allowed to drip into bilges. Renewal of the pre-filter, if fitted, is easy, but there are sometimes filters tucked away where you might not expect them. Usually there is one fitted in the fuel lift pump if it is of the diaphragm type, and there may well be one in the carburettor itself. Check with the owner's manual.

When a pre-filter is fitted, there is little chance of the main filters getting into a serious condition, but if a lot of water has been accumulating in a pre-filter of the type that will separate out water, check the whole length of fuel line and the engine filters, as droplets may have accumulated on these too.

A regular inspection of every inch of the fuel line installation should be undertaken at fitting out time before the engine is started. Copper pipe hardens with age and will, unless really well supported throughout its length, fatigue and crack; pipe clips at 200mm (8in) spacing are satisfactory. With any type of engine mounting there is both starting and vibration movement that makes it vital for the metallic fuel lines to be connected to the engine with a flexible braided metallic fuel line; check that the joint to the solid line is supported so that any movement does not reach back to the solid section and fatigue it. There should be no risk of the flexible section drooping down onto a moving part of the engine such as the alternator or drive belt. It is good practice for fuel lines to be installed above the level of the tank top so that should a fracture occur, fuel will not be continuously siphoned out of the tank; if they run below tank level they should be fitted with an anti-siphon device or an electrically-operated fuel shut off valve. Solenoid operated shut off valves are, in any fuel installation, a worthwhile safety feature. Where $\frac{1}{4}$in BSP size (and upwards) piping is used, a Beazley Fire Isolating Valve is highly recommended. These valves automatically shut off the fuel supply when a fusible link reaches a specific temperature. They are also available with Multi-Facility Adaptor switches which allow remote shut-off, fire alarm or shut down for a broken pipe, excessively high or low pressure. Where fire in boats is concerned, prevention is better than cure.

Air and Air Cleaners

Because air is free and the engine uses a great deal, it is taken for granted. Some owners try to eliminate it altogether from engine spaces so that the power unit almost chokes to death, and will certainly never run at full power. If you starve the engine of air, complete combustion cannot take place and partially burnt fuel will be ejected with the rest of the exhaust gases. Pollution is a dirty word and, when world fuel supplies are critical, it is surely sensible to see that all engines operate at maximum efficiency.

Fig 34 shows the dry paper element air cleaner fitted on the BMW marine engine, which is typical of its kind. From practical experience, I have found that the renewal

Fig 34. The air cleaner is a replaceable element in the mechanic's hand; note the inner metal flame trap of the housing.

period for the element varies enormously, depending on the amount of dust in the air where the boat is lying, or the type of boating you do. A boat that is regularly run up the beach or kept in a marina where there is a lot of wind-borne dust, must have the filter renewed every season at least, whereas in a temperate climate, the filter should last three seasons. Take a look at regular intervals and work out your own renewal dates.

In the same figure, note the perforated flame trap in the middle forming the lower part of the air cleaner unit. Many engines do not have a paper element, but every single one should be fitted with a flame trap of some kind. This is needed to douse the flame from a backfire so that it is impossible for it to ignite fuel gases outside the unit. Although it is a temptation to tune the carburettor without the air cleaner or flame trap in position, because of easier access, it is extremely foolish to do so. You risk fire or explosion, and the carburettor in any case will not be tuned to peak performance. The reason is is that even in the best air cleaner there is a little restriction of air flow when the air cleaner and flame trap are in position. Wherever possible, tuning must be done in conditions that resemble normal running conditions.

On some engines there is a wire wool air cleaner element. This should be cleaned in paraffin or kerosene and then re-soaked in oil which, on the strands of wire, catches the dust and also acts as a flame trap; this wire wool element is best left to drain down before replacement. Incidentally, the air cleaner absorbs a fair amount of air induction noise and on older engines not fitted with one, the modification is well worth looking into.

On many engines you will notice a rubber pipe connecting the rocker cover to the air intake. This is the simplest form of crankcase breather, and enables the air cleaner to breathe in vapour from the hot oils in the crankcase and, if it were not for this, a nasty oily smell would be emitted. If these oil mists and vapours were allowed to build up under pressure, they could lead to a violent explosion, ignition being caused by very hot gases blowing by the piston from the cylinder. Since piston rings are not a perfect fit, combustible gases are also drawn down into the crankcase, especially when a lot of slow, cool running takes place, where they dilute the oil and, if they build up, could cause an explosion in the crankcase. The heat built up with fast running will soon evaporate any petrol in the oil, but it must be correctly ventilated back into the engine.

In the quest for clean air, emission control systems have become very much more sophisticated than the single breather pipe.

Ford have produced both closed and semi-closed ventilation systems on UK and USA basic power units, and you should find out the servicing requirements from the manual. Watermota power units based on UK Ford 2261 and 2264 engines have a semi-closed positive ventilation system. The breather is incorporated in the oil filler cap, and there is an oil separator located close to the fuel pump which also has a crankcase emission valve fitted. The piping linking the units must be kept in good condition and any build up of dirt prevented. The emission valve is disconnected from the rubber pipe linking it to the induction manifold and pulled out from its rubber grommet to be dismantled and cleaned with petrol before replacing.

The big Ford V-8s have a positive crank-case ventilation system. Air is taken from the carburettor air cleaner through a hose to the oil filler cap on the front of the V-8's left rocker arm cover. The incoming air is cleaned as it passes through a filter in the oil filler cap. The restriction of air by the hose and the filler enable a slight vacuum to be maintained in the crankcase, as the filler cap is also sealed so as not to allow in air from the atmosphere. The air then sweeps all fumes from combustion products along through both banks of the crankcase before it enters a spring-loaded regulator valve on the top rear section of the right bank rocker arm cover. The valve is a jiggle pin type which governs the amount of air to meet changing operating conditions. The air then travels to the intake manifold through the crankcase vent hose and the carburettor spacer passage. The jiggle valve operates by sensing the intake manifold vacuum through the carburettor spacer passage and crankcase vent hose. At idling speeds, the intake manifold vacuum is high, overcoming the pressure of the valve spring. The valve jiggle pin moves to the low speed operation position where a minimum of ventilating air flow passes between the valve pin and the outlet port in the valve body. Since manifold vacuum decreases as the engine speeds up, the spring forces the pin to the full open position to increase the ventilating air flow. The return air flow is ingested into the engine as part of the mixture, so a blocked valve or restricted piping will affect smooth slow running. On no account should the crankcase ventilation system be disconnected, as this affects fuel economy and the life expectancy of the engine. Components of the system can be removed and cleaned to remove blockage or sticking with oily sludge, but the air intake test requires a special tool and this job is best left to the expert.

Carburettors

Regard the carburettor or carburettors as the engine's heart. The average person does not consider himself sufficiently skilled to do a heart transplant. Neither should the owner try to do adjustments or repairs other than jobs specified in the manufacturer's handbook. Some engine makers are so against any amateur interference that they give very little information at all about carburettors. For instance, one handbook merely says that the engine uses a Solex carburettor, which could be any one of several different types produced over the years. Confusion often occurs in carburettor identification, as engine builders tend to change carburettor manufacturers or types and put them on otherwise identical motors. This is sometimes done because of improvements in design or the carburettor manufacturer is offering a

Fuel, Carburettor and Exhaust Systems

Carburettor Type	Application*	Manufacturer
Solex 4A1	BMW-6 cylinder	A. Pierburg, Autogeratbau K.G., 404, Neuss, Dusseldorfer St. 232 West Germany
Solex 44PHN3	Volvo Penta 140 & 145 BMW-4 cylinder	La Société des Carburateurs Solex, 19 Rue Lavoisier B.P. 214
Solex 26NV	Stuart Turner Newage Lyon	92002, Nanterra Cedex France
Solex 44PA1	Volvo Penta 115/100, 130/270, 170/270	
Rochester 2GE	Volvo Penta AQ200, OMC 140	Rochester Products Division 1000 Lexington Avenue
29C 2GV Quadrajet	Mercury OMC 260 Volvo Penta 225, 260, 290	Rochester, New York, USA
Holley 6317-1 6407 & 6576 4160 & 4150	USA based Ford units Some Mercury Some Volvo Penta	Holley Carburetor Division 11955 East Nine Mile Rd Warren, Michigan 48090, USA
Tillotson HL series	Vire	Tillotson Manufacturing Co. 761-9 Berden Avenue Toledo 12, Ohio, USA
Zenith-Stromberg 175CD VN series Solex 30AHG	Volvo Penta AQ B18 Volvo MB18B & BB100 Volvo MB18F & BB30 Wortham Blake	Zenith Carburettor Co. Ltd. Honeypot Lane, Stanmore Middlesex, HA7 1EG, UK
S.U. HD6	**Marinised Rover V8**	S.U. Carburettor Co. Ltd. Erdington, Birmingham, U.K.
Motorcraft 1V	UK Ford 2260/2270 Marinised units	Ford Motor Co. Ltd. Industrial Power Products Arisdale Ave., South Ockenden, Essex, RM15 5TJ, U.K.

Application Guide: Manufacturers reserve the right to change specifications and this list does not constitute a precise guide as to which carburettor was, or is, being used on any engine ever manufactured. However, carburettors are usually engraved and the address list should enable an owner to track down the maker either, initially, through the engine builder, or the carburettor manufacturer.

Table 4. Carburettor guide to manufacturers
This table is not meant to be comprehensive either as a guide to every manufacturer or to cover every marine engine. It will however serve as a basic guide to those seeking further information

cheaper model that will do the same job. Table 4 on **page 67** lists the addresses of carburettor manufacturers, though your first contact for information should be the engine manufacturer.

There are certainly some 'dont's' regarding care of carburettors. They are extremely reliable instruments and should not be pulled apart at the first sign of engine trouble. In fact, unless you are completely familiar with them, they should not be pulled apart at all. If you do have to clean a jet or passageway, as per maker's instructions, don't prod about with wire as this will scour fine drillings and ruin them. Finally, see that the gaskets are perfectly lined up and don't improvise with liquid gasket sealants, lavishly applied. I once saw a carburettor that had been treated so extravagantly with gasket goo that many of its thirty or so fine drillings were completely blocked up, thereby ruining an expensive item.

Gaskets are important for keeping extra air out, so as not to lean the mixture, and fuel in because a fuel leak will probably cause a fire. The maker's gaskets should always be used, as the material is chosen for specific qualities. For example, he may specify a heat isolating gasket between the manifold and the carburettor so that heat is not transferred. The gasket may, however, be needed to do exactly the opposite, as the heat can be utilised to help vaporise fuel. Lining up a gasket perfectly so that it does not overlap and partially **block passage ways is good practice.** Any

Fig 35. The Solex 4A1 Carburettor on the BMW 6-cylinder marine engine.

A) The TN starter; automatically produces rich starting mixture. B) Starter body incorporating bi-metallic spring and electric heating which control part of cold starting choke procedure.

Note finger pointing to small connector for rubber tube; the amateur can check all such tubes and see that electrical wires are properly connected; otherwise this complex carburettor should not be touched by the amateur.

gasket will do less than its job if the mating surfaces have not been cleaned or if the proper torque is not applied to the nuts or bolts holding it in place. You can buy gasket papers by the sheet and, by placing them over the casting and gently using an engineer's ball-pein hammer, tap a new one out. Then, tapping on the edge of the casting, cut out the new gasket.

Now, a look at one of the most advanced carburettors made today.

Solex 4A1. Fig 35 shows the BMW six cylinder marine engine with the Solex 4A1 carburettor. Illustrated are some of the details that you can look for if there are problems away from a service station. The finger points to a small bypass air pipe. On this particular carburettor, and many similar, there are a number of pipes connected other than that for fuel. Before you attempt any work, know what each of these pipes does and, even when you are not familiar with their function, you can see that each is connected properly and not perished or broken. You already know that the crankcase breather may be one of the larger rubber pipes, and you may also find a water pipe that controls the thermostat on the carburettor which, in turn, controls part of the automatic cold starting device. This lets extra fuel flow to the engine when it is cold without allowing it to over-choke on too rich a mixture when it is started hot, or as the engine fires up. The thermostat is also linked to an electrically operated cold starting device which combines its function with the fresh water cooling system.

Another pipe you may come across is a thin metal vacuum pipe. This is usually connected into the carburettor manifold to sense vacuum depression and operate the automatic advance and retard on the ignition. You will find this pipe running to the side of the distributor body where it connects into a diaphragm housing. It can become cracked through careless handling or fatigue, so ruining the automatic ignition timing.

And, now, the one pipe you expect to be going to the carburettor – the fuel pipe. Many carburettors have their own filter where the fuel pipe enters the body. On the BMW Solex 4A1 this is immediately behind the banjo where the pipe enters the body of the carburettor. Your engine manual should tell you if there is one on your unit; they are an extra safeguard against dirt getting into the precision works. They usually consist of a small cylinder of extra fine copper gauze. To clean it, remove it from the body and wash in petrol. Immediately below the carburettor you will also find an in-line fuel filter – a disposable and renewable item for pre-filtering the petrol.

Your curiosity may be aroused when, on modern carburettors, you find plastic plugs blocking holes. The 4A1 has a couple, and you will see one as you face the manifold of the engine at the bottom of the unit. These plugs cover the idle mixture control screws and are tamper-proof devices. Some bureaucrat sits somewhere working out what colour plastic plugs the manufacturer should use so that they conform to a particular colour code. With the 4A1, if your BMW is not idling at the suggested speed (800 to 900 rpm) then you are allowed to adjust the throttle shutter stop screw (found on the right hand side of the unit facing the manifold) a quarter turn either way. If more than this is needed, the makers suggest that the unit should be properly serviced so that the two mixture screws can be adjusted professionally. It is safe to adjust the idling speed on all makes of carburettors, with similar rotary limits of the idle speed adjusting screw. More extreme adjustment means that the engine is getting the wrong mixture. Only on simpler carburettors, as you will see later, should you adjust the idling mixture screw.

An additional complication on the more advanced carburettors such as the 4A1 and the Rochester 2GV are electrical connections. Electric chokes are operated by bi-metallic strips heated by an electrical resistor. Poor starting may simply be a case of a wire adrift or a dirty connection.

From the foregoing it will be apparent that the twin choke carburettor is not the province of the amateur, but there are a number of visual inspections and one or two small jobs that you can do to keep the units in good order.

One feature common to any carburettor is the need to keep fuel at a precisely determined level. This is, for the most part, achieved by having a chamber containing a float. As the float moves upwards as the chamber fills, it shuts off the supply until the level drops. There are variations in the methods of shut-off: some floats work through a lever and some by a vertical pin in the float which closes the orifice in the top of the chamber. In old designs there was a serious fire or explosion hazard because the needle did not shut off and the carburettor flooded. Today most carburettors are of the safer downdraft type, such as Holley, Rochester, Solex 4A1 and Webber, to name just a few. The big advantage of these is that, if the float chamber floods the fuel only runs harmlessly into the engine. US Coast Guard rules are worth noting, and they are specific about the danger of flooding petrol from the carburettor.

Each carburettor must not leak more than five cubic centimetres of fuel in thirty seconds:
1. When the float is open;
2. The carburettor is at half throttle; and
3. The engine is cranked without starting; or
4. When the fuel pump is delivering the maximum pressure specified by its manufacturer.

Each updraft and horizontal draft carburettor must have a device that:
1. Collects and holds fuel that flows out of the carburettor venturi section towards the air intake;
2. Prevents collected fuel from being carried out of the carburettor assembly by the shock wave of a backfire or by reverse air flow; and
3. Returns collected fuel to the engine induc-

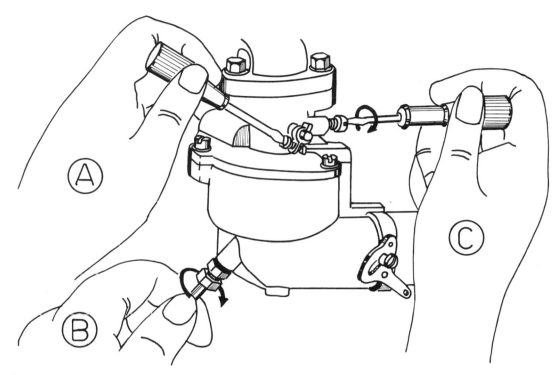

Fig 36. Adjustment of a simple carburettor.
A) Idle speed adjusting screw. B) Idle needle valve.
C) Main jet needle valve.

tion system after the engine starts.

The simplest carburettor, the Villiers B10/1 shown in Fig 36, sometimes found on small two- and four-stroke engines, does not have the sophistication of the Solex 4A1 but is perfectly effective for simpler designs. Basic components are a float chamber, a main running jet (B) and an idle jet (C). Adjustment of fuel is made by means of needle valves. As the name implies, the fuel moving through the orifice is controlled by the taper on a needle, which is an extension of a threaded section; the needle is turned by either a screw slot or knurled cheese head section. Damage

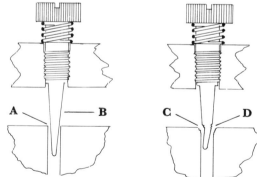

Fig 37. Carburettor Needle Valve–use and abuse.
A) The jet orifice must have a clean edge. B) The needle must be perfectly straight with original taper. C) Overtightening or abuse with a wire has damaged the orifice. D) Overtightening has damaged the needle taper and it is bent; satisfactory operation is now impossible.

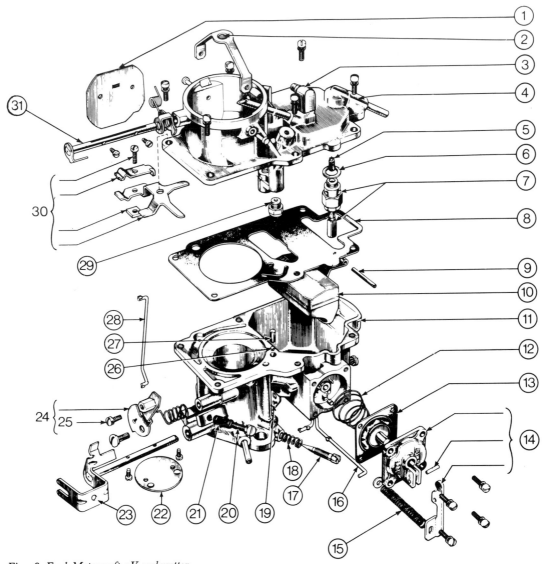

Fig 38. Ford Motorcraft 1V carburettor.
1) Choke plate. 2) Retainer. 3) Elbow. 4) Upper body. 5) Filter. 6) Gasket. 7) Needle valve assembly. 8) Gasket.
9) Float hinge pin. 10) Float. 11) Lower body. 12) Accelerator pump spring. 13) Diaphragm. 14) Accelerator cover
assembly. 15) Accelerator pump over-travel spring. 16) Accelerator pump link rod. 17) Idler screw. 18) Spring. 19)
Accelerator pump lever. 20) Throttle idler screw. 21) Spring. 22) Throttle plate. 23) Throttle shaft. 24) Cam assembly.
25) Screw. 26) Discharge ball. 27) Weight. 28) Choke plate cam link. 29) Main jet. 30) Choke bracket assembly.
31) Choke shaft assembly.

is done if it is screwed down too tightly into the orifice to start adjustment from a closed position: the needle becomes burred on the edge of the orifice, the soft metal of the casting is damaged and you will never get a smooth adjustment again without replacing the casting and the needle (Fig 37).

On simple carburettors there are usually two needle valves controlling the slow running or idle jet and the main jet. If you don't have technical details, each should be opened about one-and-a-half turns from the closed position as a rough starting point, while the engine is warmed up. If the engine won't start, check if it is flooding (raw fuel on the spark plug) or is too lean (no smell of fuel in the cylinder–plug dry); increase or decrease supply from the main jet accordingly. Check the float chamber to see that fuel is flowing in correctly. Now, turn in the main jet needle valve to lean the mixture until the engine begins to misfire. Note how much you have turned it by looking at the relative position of the screw slot or a mark on the knurling, and then enrich the mixture by turning the needle valve out until the engine runs unevenly. Adjust the needle valve to exactly half way between these two positions. Set the idle jet by fixing the throttle in the idle position and the engine running at the specified idle speed. A slow idle should be around 700-800 rpm on small engines, and a fast idle around 1,400-1,700 rpm. The idle needle valve is now adjusted with the throttle held in position, enriching and leaning until the

best running is obtained. The final test is to see that the engine accelerates without missing. If, on opening the throttle, the engine coughs or lags for a moment, open the main jet needle valve slightly to give a richer mixture. When the engine accelerates, the air rushes in more easily than the extra petrol, and an over-lean mixture is produced. On larger engines this hesitation would not do at all, and such a rough enriching device would be inefficient and uneconomical. With fixed choke carburettors a pick-up device is pretty well essential to give satisfactory acceleration from low speeds. This section of the carburettor delivers an extra drop of petrol into the choke when the throttle is suddenly opened, and it is called an accelerating pump.

For a last look at the fixed choke type of carburettor, I have selected the Ford Motorcraft 1V model which is shown exploded in Fig 38. This, like other carburettor illustrations I have used, should only be regarded as an example, as variations exist between models in specific applications.

On the lower right hand side of the figure you will see the accelerator pump I have just referred to. This is a relatively simple unit compared with the Solex 4A1, so you will appreciate why certain carburettors should not be tampered with. There is more than enough for the amateur to deal with in this one.

The main problem likely to be encountered is dirt which settles in the bottom of the

Fuel, Carburettor and Exhaust Systems

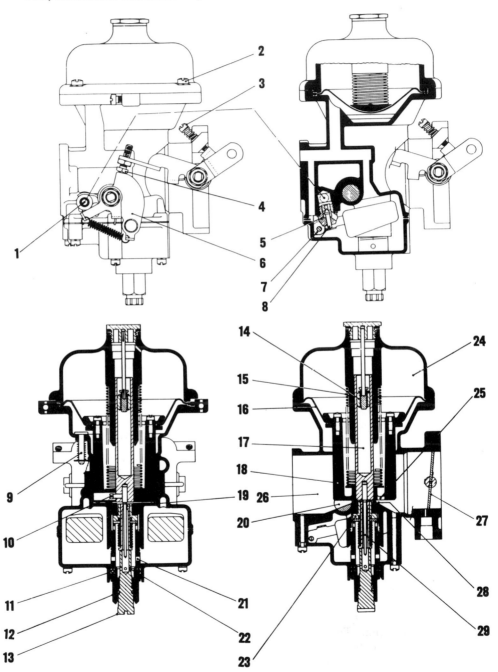

float chamber. Even with the best pre-filtering in the world, this gradually builds up and, if not removed, blocks the float chamber intake or, worse, gets into the jets. Each season the float chamber should be cleaned according to the Ford Service Manual for the 2260 and 2270. I might say that the Ford publication is greatly superior to many other marine engine makers' handbooks.

The second type of carburettor is the constant vacuum, typified by the Stromberg CD series model shown in Fig 39. Because of the unrestricted air flow directly through the bore (26), there is good top end performance at comparatively little manufacturing cost compared to the twin fixed choke types, which have overall better performance. However, in marine applications, there is much to be said for simplicity.

There is, for example, no separate idling circuit on this model. The fuel comes through the jet orifice (19), the amount being controlled by the setting of the adjusting screw (13) which may be screwed clockwise, with a small coin, to decrease mixture strength, or anti-clockwise to enrich the mixture. The metering needle is a precision component, the perfect taper being made so that as it is lifted in the orifice by the air valve (18), the increasing volume of fuel matches the increasing volume of air. Apart from keeping the needle straight, it must bed perfectly within the jet orifice. A danger warning: testing can only be done when the flame trap/air cleaner is removed by raising the air valve with the lifting pin (9) and checking that it falls back freely. If it is not centred proceed as follows:

1. Lift the air valve (18) and tighten the jet assembly (12) fully.
2. Screw up the orifice adjuster until the top of the orifice (19) is just above the bridge (28).
3. Slacken off the whole jet assembly (12) approximately half a turn to release the orifice bush (23).
4. Allow the air valve (18) to fall; the needle will then enter the orifice and automatically centre it. You can assist the air valve to drop by pushing it down gently with a soft metal

Fig 39. The Stromberg Series CD carburettor.
1) Petrol inlet. 2) Screw and spring washer for fixing cover. 3) Throttle stop screw. 4) Fast idle stop screw. 5) Fuel inlet needle valve seating. 6) Starter bar lever-operated by cable from instrument panel. 7) Twin expanded rubber floats on a common arm. 8) Needle that controls fuel inlet to float chamber. 9) Air valve lifting pin. 10) Screw holding tapered metering needle in place. 11) 'O' Ring. 12) Bushing retaining screw. 13) Orifice adjusting screw. 14) Damper assembly. 15) Coil spring to assist downward movement of air valve. 16) Diaphragm–check for rupturing. 17) Hollow guide rod housing dashpot or hydraulic damper; top up. 18) Air valve; check for smooth up and down movement. 19) Jet orifice. 20) Starter bar. 21 and 22) Holes connecting float chamber with jet orifice. 23) Bushing for jet orifice. 24) Chamber. 25) Drilling in air valve to transfer manifold depression to chamber. 26) Bore. 27) Butterfly throttle. 28) Bridge of throttle bore. 29) Metering needle.

rod via the damper access at the top.

5. Tighten the assembly (12) slowly, checking frequently that the needle remains free in the orifice by raising the air valve about $\frac{1}{4}$in and letting it fall back freely. The piston should then stop firmly on the bridge.

Now you must set the idle. Two items have to be adjusted, the throttle stop screw (3) and the jet adjusting screw (13). With the air cleaner and damper removed, the air valve (18) is held down on the bridge of the throttle bore (28). The jet (13) is now screwed up until it just comes into contact with the underside of the air valve. The jet adjuster is then screwed back three full turns to establish an approximate position. Run the engine until it is warm and then, by means of the throttle stop screw (3), get an idle speed of about 600-700 rpm; the mixture will be correct when the engine beat is smooth and regular, and by gradual adjustment of the screw (13) the final position will be determined. As a check, a thin screwdriver ($\frac{1}{32}$in) can be used to lift the air valve. If the engine speeds up appreciably the mixture is too rich, and if it stops, too weak. Correctly adjusted, the engine speed should remain fairly constant, or fall slightly when the air valve is lifted. Many of the higher performance engines have twin carburettors of this type and I would suggest that the tuning is best left to the professional, as it takes a great deal of skill to get them perfectly synchronised.

The dash pot (17) is normally topped up to within $\frac{1}{4}$in of the top with normal clean engine oil, about SAE 20 or as recommended by the engine manufacturer. In caring for these carburettors, it is always recommended that the throttle is left in neutral and the manual choke pulled out fully for starting purposes. This is because the choke already operates the starter bar (6) which lifts the air valve and with it the needle to enrich the mixture and, simultaneously, a cam on the bar opens the throttle beyond the normal idle position according to the setting of the fast idle stop screw (4) to give a fast idle when the motor is cold. See that the choke is returned to the closed position gradually as the engine warms up, since it will not appreciate the over-rich mixture.

Although, sadly, the small two-stroke engine has seen a decline in recent years, Valmet of Finland who make the successful two-stroke Vire, have found a steady market. It is a compact and well designed unit producing some 7hp at 143lbs of weight (Fig 40). The earlier model produced 6hp and used one of the simple Solex carburettors which had the float needle chamber orifice reduced to prevent flooding experienced in earlier models. The trouble was caused by fuel pump pressure which overcame the resistance produced by the float on the float chamber needle. Of particular interest is the fact that it uses a Tillotson HL series diaphragm carburettor. This is an all position carburettor which was originally developed for the chainsaw industry–and you can get

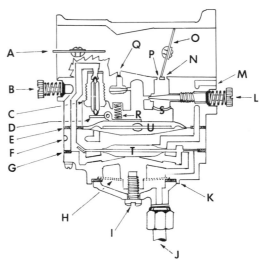

Fig 40. The beautifully compact Vire two-stroke marine engine.

Fig 41. The Tillotson HL Series Diaphragm carburettor. A) Choke shutter; close for starting. B) Main jet adjusting screw. C) Seat for inlet needle with inlet needle resting on D. D) Inlet control lever moving on fulcrum pin, tensioned by R. E) Diaphragm cover gasket. F) Atmospheric vent hole to chamber on lower side of diaphragm. G) Fuel pump gasket. H) Fuel filter screen; must be kept clean. I) Filter cover retaining screw. J) Flexible fuel inlet pipe. K) Filter cover gasket. L) Idle adjustment screw. M) Impulse channel connected to fuel chamber (upper part of diaphragm (T) activates it to pump fuel by engine pulsation. N) Primary idle discharge port. O) Throttle shutter. P) Secondary idle discharge port. Q) Main fuel discharge port. R) Inlet tension spring. S) Welch plug; badly clogged idle fuel supply may necessitate its careful removal and replacement. T) Fuel pump diaphragm; below diaphragm is fuel chamber and above the pulse chamber. U) Diaphragm; upper chamber controlling fuel inlet and lower chamber atmospheric; check diaphragms for puncture or age hardening.

information and servicing from that source if you need it. I first became acquainted with these units when I was go-karting in the Far East. They were ideal when considerable 'G' forces were applied on sharp bends, as they always kept the fuel flowing to the engine–just as with chain-saws which are used in all sorts of awkward positions. Since boats are prone to being thrown around by waves, this is a welcome attribute.

Looking at Fig 41 you see that the lower part of the carburettor consists of a fuel filter (H) which should be regularly cleaned. Above this is the integral fuel pump, actuated by the diaphragm (T) which gets its movement from the pulsations of the engine. The fuel inlet needle is housed in an insert (C) and rests on the inlet control lever (D). The inlet valve is kept closed by the tension of the spring (R) acting through the lever. The metering diaphragm (U) is subject to engine suction on the upper metering chamber side and to atmospheric pressure on the lower vented side. The atmospheric pressure on the vented side pushes the diaphragm towards the inlet control lever, opening the fuel inlet needle to allow fuel to enter the metering chamber. On the Vire the carburettor metering system includes a ball check valve (Q) in the main nozzle. This allows fuel to flow into the mixing passages and prevents air from entering into the metering chamber. Since it is engine and atmospheric pressure that works this type of carburettor, it is of prime importance for the diaphragms to be in perfect condition, and the gaskets must give nothing less than a perfect seal. Diaphragms seldom puncture but they do age and harden, so a rebuild is to be recommended every third season or so.

Filthy fuel will, as in any carburettor, be a major problem. The internal passageways become blocked and the welch plug (S), which is similar in engineering design to a core plug, should be carefully cente drilled and pried out so that the internal supply channel and discharge ports can be cleaned; use a ⅛in diameter drill and ensure that the plug is only just pierced, so avoiding damage to the body casting. When replacing the plug,

make it fit perfectly in the casting shoulder, gently flattening it into place so that it is just tight and no more.

As a starting point for tuning this carburettor, open the main jet one and a quarter turns and the idle jet about three quarters of a turn. Close the choke (A) and open the throttle shutter (O) slightly. Remember that the fuel pump needs a little more time to prime than other types but, as soon as the engine fires up, open the choke slightly and idle the engine until it is warm. Now, with a fully open choke and a completely closed throttle shutter, readjust the idle adjustment screw (L) until the engine is running–at about 1,200 rpm in the case of the Vire. The throttle shutter may have to be adjusted by means of the external idle speed screw, so that the engine rpm does not drop below 1,000 rpm when in gear. When restarting a warm engine using this type of carburettor, you should not use the choke; if there is poor acceleration, enrich the idle mixture. Adjustment of the main jet for full power is, say Vire, best done with the boat under way. I prefer to make an extra strong set of mooring lines ashore and, with plenty of fenders in the right place, test the engine under power. There is always the possibility of the engine dying on you. After a considerable run, take the plug out and check its condition to see whether the mixture is right.

At laying up time, carburettors appreciate a basic clean out of filters, and the Tillotson will appreciate having the fuel/oil mixture flushed out. Remove the spark plug and earth the lead so you can safely turn the engine over. Then, with a drip can below the carburettor body to catch fuel, and the two screws (B) and (L) carefully removed, let the fuel flow out and then drain down. Replace the two screws.

Fuel Pumps

In the past, fuel was delivered to the engine by gravity feed. I hope you won't find such a system on a marine petrol engine today, for it constitutes a fire hazard if any small leak occurs. It could flood the boat as well as the carburettor. Some boats that were not fitted with gravity feed had small electric pumps to lift the fuel to the carburettor, but these may be regarded as obsolete as well. The mechanical diaphragm pump now finds almost universal approval, as it is extremely reliable and will deliver the correct volume of petrol safely to the carburettor at the correct pressure without much attention from the owner. However, you do need to see that high or low pressure from the pump is not caused by other faults in the fuel system; air leakage in the fuel line can cause low volume and pressure. A restricted fuel tank vent will do the same, and there is even the danger that severe depression will cause fuel lines or the fuel tank to collapse (the atmospheric pressure is over 14lb per square inch, which is quite sufficient to cause this kind of

trouble); low pressure fuel will affect the engine's performance by starving the carburettor, leading it to produce a lean mixture; excessive pressure will cause flooding and high fuel consumption. Pressure and volume tests are jobs for the professional, but you should be able to tell him what you suspect.

Some fuel pumps, such as the double diaphragm used on some Mercury power units, are not repairable. These pumps incorporate a sight glass so that you can see immediately if a fault has developed. The materials of the two flexing diaphragms have different characteristics so it is unlikely that the second safety diaphragm will fracture at the same time or before the primary one.

Fig 42 shows a typical diaphragm fuel pump as used on the Ford 2260 and 2270 marine engines. The lower body casting (Q) houses the operating lever (N) and diaphragm (H) and has a mounting flange to attach the pump to the engine. The operating lever has a spring loaded rocker arm that bears against the camshaft eccentric and a link connected to the diaphragm pull rod. Rocker action from the cam only pulls the diaphragm downwards and it is returned by the spring (T). As the link (L) is separate from the rocker arm, it allows the pump to free wheel when the carburettor chamber is full. These pumps are reliable but need basic servicing each season. The filter bowl should be removed and cleaned along with the filter screen (D). The bowl gasket (C) should be kept in

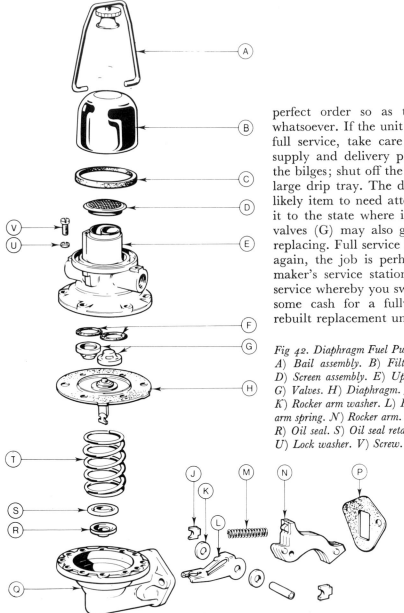

perfect order so as to allow no leakage whatsoever. If the unit is to be taken off for a full service, take care to see that the fuel supply and delivery pipes do not leak into the bilges; shut off the supply and provide a large drip tray. The diaphragm is the most likely item to need attention as age hardens it to the state where it might rupture. The valves (G) may also get jammed and need replacing. Full service kits are available, but again, the job is perhaps best done at the maker's service station. Many firms run a service whereby you swap your old unit plus some cash for a fully guaranteed factory rebuilt replacement unit. If you do take out

Fig 42. Diaphragm Fuel Pump.
A) Bail assembly. B) Filter bowl. C) Bowl gasket. D) Screen assembly. E) Upper body. F) Valve gasket. G) Valves. H) Diaphragm. J) Rocker arm pin retainer. K) Rocker arm washer. L) Rocker arm link. M) Rocker arm spring. N) Rocker arm. P) Gasket. Q) Lower body. R) Oil seal. S) Oil seal retainer. T) Diaphragm spring. U) Lock washer. V) Screw.

the diaphragm, make sure you mark the position of the diaphragm tab on the side of the body before you move the upper body (E). This is so that when you put it back it goes the same way and the two slots engage in the rocker arm link (L). The diaphragm is twisted a quarter turn to disengage the rod from the link. The valves on this model are staked to retain them, but other types have a retaining plate and screws. The rocker arm spring (M) and diaphragm spring (T) may weaken with age and need replacing; eventually, wear will occur on the rocker arm pin. The gasket (P) is important, for it provides the correct positioning of the rocker on the cam and isolates the pump from the heat of the main engine casting. The correct thickness gasket thus ensures there is no mechanical stress on either cam or rocker and that vapour locks are eliminated in this part of the system. The pump makers usually offer a full servicing kit, but I always carry one new replacement when I go to sea, as servicing the unit is a fiddly business, only to be undertaken in a clean workshop and not in a seaway.

Exhaust Systems

The high temperature in the exhaust system is the prime cause of serious corrosion problems. Dissolved gases produce chemical attack; together with water flow impingement attack and continual immersion in salt water it makes for an unwholesome mixture. The main problem for the engine designer is to ensure that the power unit can exhaust its burnt gases as easily as it breathes in air. This sounds fairly simple but, in the confined and enclosed space of the engine area, there are many practical problems to overcome—that is, for installations other than outdrive power units which have their own integral exhaust system. The engine builder depends almost entirely on the boat builder following recommended practices. Back pressures should be contained within acceptable limits, while cooling water or the water on which the boat is floating, must not be able to get back into the engine. The system must also silence the exhaust to an acceptable degree and clear obnoxious fumes. Without complete integrity of the exhaust line, carbon monoxide could escape into the saloon and kill. Since sections of the exhaust on both inboard and outdrive power units are at, near or below the waterline, a fault could cause the boat to sink. Consequently, exhaust systems should be inspected at least twice every season.

Wet Exhaust Systems. Fig 43 shows the typical components of a wet exhaust system where, as stated above, design is critical as regards its relation to the waterline. To eliminate back pressure, the best installation is a straight-through run with a constant fall from the water injection bend. All deep sags in pipework (E) allow water to settle, and I'm sure that the humidity this creates can be a source

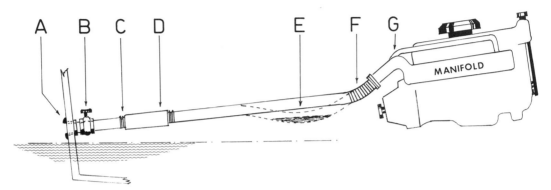

Fig 43. Inspection points in the exhaust system.
A) Skin fitting. B) Gate Valve. C) Double stainless steel clipping on all rubber to metal. D) Silencer in good condition and correctly placed. E) Sags to be avoided; any section that does collect water to have drain cock and be drained after use. F) Flexible exhaust section must be provided on every installation either rubber or metal type. G) Water injection bends subject to severe corrosion.
Note: the whole length of a system must be correctly supported to avoid stress and chafe.

of corrosion for valves and cylinders that happen to be open within the engine. If swan necks or sags are necessary, each section where water is likely to accumulate should be fitted with a drain cock. The problem with these is that you might forget to drain them down when you leave the boat because, so often, they are awkward to reach. The water injection bend (G) is supplied with raw water from the engine after it has done its cooling work there, and is then sprayed on the hot exhaust to cool it. Always anticipate corrosion opposite the injection pipe because of impingement attack aided by dissolved chemicals from the exhaust gases. Design of these bends has improved in recent years, altering the angle of the spray so that the

full force is not taken on a small area of the pipe wall. However, it is recommended that an internal inspection should be given every third year by removing the unit completely from the manifold. It sounds crude, but a simple test is to poke the most vulnerable area with a sharp steel spike from time to time. It is better for the spike to go through a weak spot when you can deal with the situation, rather than find the engine room flooding at sea.

The outlet should always be fitted with a gate valve. Builders often omit this item to save money yet, if the inboard side of the exhaust should fail, the boat would sink if there were no means of shutting off the outlet opening. Even when the outlet is above the waterline, waves slapping against the transom can soon fill a boat if the exhaust line is faulty. Both gate valve and transom fittings (A) need to be regularly inspected for service-ability and corrosion.

Pipe materials vary widely: stainless steel, copper, synthetic rubber and iron are all used. Each has its own merits and draw-backs, and it is as well to know something about them all so that you can make a con-sidered choice when it comes to replacement.

Do not mix metals in an exhaust system. Curves, even when they are flexible ones, should be as smooth as possible so as not to obstruct the flow of gases, and not to exert excessive stretch on the inside of the pipe bend; this weakens the metal by thinning it and putting a stress pattern into it which promotes eventual disintegration. The one exception to mixing metals is where rigid pipework joins the engine manifold. To absorb vibration, there should be a length of flexible hose inserted into the system near the manifold; these are usually made of stainless steel. It is also sensible to assemble all metal systems with short lengths of rubber exhaust hose clipped over joints, so that you can dismantle them easily and test metal thickness from time to time.

Stainless Steel. This is a fine material except when permanently immersed in sea water. It is then liable to pitting attack, and is not ductile like the copper or copper-bronze which found favour in the past.

Copper. Highly polished copper pipes looked well when they swept through engine spaces. They were filled with resin for bending pur-poses, and the skilled smith made his crafts-manship a pleasure to behold–but you would probably have difficulty in finding such skilled people in boatyards these days. Copper, however, suffers serious leaching of the zinc element and, if you see it turning pink, suspect serious erosion.

Synthetic Rubber. Synthetic rubber hoses were originally developed for diesel engines with exhaust emission temperatures lower than in petrol engines. However, they are more and more used on petrol installations and, pro-vided there is a constant supply of cooling water, they should have a reasonable life expectancy. Excessive heat, though, does cook them, so that they harden and the material delaminates; look out for the outer cover starting to peel, which is an indication that things are not as they should be. All rubber exhaust pipes to the outside of the boat should be double stainless steel clipped (D) for safety.

Iron. Iron pipes are cheap for the builder but not for the owner. They corrode rapidly and are a rotten job to remove when they corrode through.

Maintenance. For some years I have had con-siderable success in treating cool sections of the exhaust system, made of the usual metals, with epoxy resin and two-can polyurethane paint systems. On the whole I have found the epoxy systems first class, with negligible corrosion after five years service. The polyurethanes seem to stand up to higher temperatures without discolouring. But protection rather than cosmetic value is more important. When the boat is left afloat, keep valves closed.

Reaching to get at the skin fitting is not a favourite pastime. It is often only accessible on sailing boats to the thinnest of the seven dwarfs. You may be able to dangle junior down there, but will he have the strength to remove the clips and pipes for inspection? If the metal is thinning, it is essential to remove the fitting even if it looks as though it will last some time. The reason is that the nut on the inside puts some stress on the main body and it might cause fracture. Overtightening is never advisable, although you might be tempted to do this when you are putting in the seacock. The silencer (D) is often omitted in the wet system where rubber exhaust hose is used, as the latter already absorbs most of the noise. However, the silencer can diminish sound level and, if it is equipped with baffles, ensure that exhaust water does not find its way back towards the power unit. There are definite positions to fit a silencer in order to get the maximum benefit, and the best place is at the exhaust manifold but, as there are problems of cooling when neoprene or other synthetics are used, the next best available places are at two fifths or four fifths of the exhaust system's total length from manifold to transom.

Example:
Total length of exhaust pipe: 15ft with a 2ft muffler to be included.
$$2/5 \times 15\text{ft} = 6\text{ft}.$$
Subtract half the length of the muffler:
$$6\text{ft} - 1\text{ft} = 5\text{ft}.$$
So, the pipe from the manifold will be 5ft, the silencer 2ft, and the tail pipe 8ft long. If, on inspection, this is a bad place to install the muffler, try the four fifths calculation.

Drumming and Chafe on parts of the exhaust system transmit noise and can increase rather than attenuate it. Solid pipes are best soft mounted on hanging brackets, and even rubber hose should be isolated from bulkheads and structural works if it is to be deterred from transmitting exhaust noise. Isolate pipes to prevent chafe; rubber hose is particularly vulnerable where it passes through bulkheads or rests on some flat structural part. My method is to fold some light alloy sheet about 18-20 swg around the pipe and clip it in position with stainless steel clips at either end. This, in turn, sometimes needs isolating through bulkheads by wrapping with a soft rubber padding material. Check annually to see that chafe gear is still in place and working correctly.

Dry Exhaust Systems are not popular on pleasure craft as they tend to be noisier, eject fumes where they are not wanted and are, in my opinion, a greater fire hazard on a petrol-engined boat. One fear at the back of my mind is fuel dripping onto a hot exhaust and vaporising. True, part of the manifold intake may be designed to help the carburettor vaporise the fuel, but that is where it is safe. On the outside a space might fill with explosive vapour and the first spark or hot spot would spell disaster. Never, then, in any circumstances have a fuel pipe run over the top of an exhaust pipe. The dry exhaust leaves the manifold and should drop into a downward loop with a drain cock at the bottom. This drain cock must receive attention after each run, as condensation from the combustion process will tend to collect at the lowest point. A silencer of the automotive type has to be part of this system and should be built in as already outlined; they are now available in stainless steel, and could be well worth their extra cost in a corrosive environment. On the first section of the exhaust riser, a flexible section is built in to take care of the engine movement. After this flexible section, the rest of the exhaust can be strongly bracketed to the bulkhead to rise through the deck, or overboard to the transom or topsides. Deck openings should have weatherproof flashings to keep rainwater out while isolating the hot exhaust pipe from deck or bulkheads. All parts of the system other than the flexible section must be well lagged to keep heat

dissipation to a safe level. Expanded metal foils wrapped round pipes and kept in place with asbestos tape are ideal, provided that the asbestos tape is of a safe non-cancer producing variety. This lagging does mean that to keep a dry system in good condition and for regular inspection, you have to remove this outer layer first.

I might mention the dangerous practice of some boatbuilders and repairers of using flexible steel, instead of stainless steel, pipes. Fleixble steel pipe corrodes so fast that it is in constant danger of leaking dangerous exhaust gases; it should be condemned and the proper material substituted. It is important to lag even a short section of a wet system in the area before cooling water is injected. I don't like asbestos tape in direct contact with pipework as it seems to hold damp, allowing corrosion to take place underneath; isolate it with metal foil. The larger engine manufacturers often have special publications with installation recommendations, while smaller concerns publish theirs with the owner's handbook. I would certainly recommend you to check that your boat installation corresponds with these recommendations.

six

The Engine's Electrical System

The weakest link in any type of engine installation is, inevitably, the electrical system; electricity and water just do not get on with each other. Even when the engine is comparatively dry, condensation and the consequent corrosion will take toll of the electrics aboard the boat. Deterioration will be worse when the original installation is below par; and lack of maintenance will deal a death blow. How can an original installation be below par? Inferior equipment, not designed or intended to be used for a marine environment, is sometimes installed by penny-pinching builders. The basic equipment supplied by the engine maker will certainly be well up to standard, but items like ancillary wiring and switches are often inferior. The boat owner sometimes adds to his own problems by adding electronic gadgets guaranteed to overload the engine's generating capacity and the batteries. With other than the simplest installation it is essential to have two banks of batteries – one for engine starting and the other for the boat's services.

It is comfortable to assume that the electrical installation has been correctly installed, but the sad truth is that it frequently leaves a great deal to be desired. Look out for wires on the thin side, runs that leave them drooping in space with poor clipping to support them. One of the most common faults with marine electrical equipment is caused by running items at a voltage lower than that intended; this is usually the result of voltage drop, which is caused by the resistance in the conductor wire (which increases in proportion to the length of its run). The answer is to put a bigger cross section conductor in the sheathing and so reduce its resistance. Call in the professional to check that the voltage at the end of power lines is sufficient for the unit it is to power. Chafe is another deadly enemy of wiring. Inspect to see that wiring does not come into rubbing contact through engine vibration. Vibration itself is a wiring destroyer as it fatigues the conductor wires, especially at terminal ends. The use of a single solid conducting wire is not good practice. I recommend multi-strand conductors in any marine installation.

Modern installations have gone over to the crimp type terminals, first used in the automotive industry. They are excellent so long as they are crimped at the correct pressure. The chap with the vice-like grip often overcrimps them with the result that some of the conductor wires are cut, and the rest are put under such pressure that with a little vibration they will fatigue and break. Most engines these days are supplied with a wiring loom. Although you must check the terminals, what happens to the wires inside can be a source of embarrassment. You might expect the people who make looms to use whole lengths of wire, but it is not uncommon for bits to be joined together well out of sight; when a breakage occurs it is a very expensive job for a professional to trace the fault. It may well be cheaper to replace the whole loom rather than repair the old.

Fig 44 shows the basic ignition circuit for a six cylinder petrol engine. Each part of the system needs regular visual checks and specific servicing if it is to perform satisfactorily.

Starting Batteries

These are usually of the lead acid type. In standard designs maintenance means keeping the electrolyte level just above the plates by adding distilled water. Leaving exposed plates dry will damage them; overfilling may well cause acid to be spilled and result in damage to the boat or yourself. Two useful items for battery servicing are a battery hydrometer and a voltage meter, such as the one Smiths make, which is connected to the battery with crocodile clips and, after forty seconds or so, will give a true reading of the battery voltage. Many boats now have built-in voltmeters and they give most useful information on the state of the batteries all the time. During charging when the engines are running, you might expect up to 14 volts or so pushed into the batteries, while a heavy load on the service batteries, as when a craft is in harbour, will soon show the voltage dropping below 12. Low voltage must be seen to if equipment is not to be damaged by under-volting. During lay-up this type of battery must be charged regularly. Normally, the lead acid battery will slowly lose its charge whilst standing and, if you make good this loss at six weekly intervals, it will remain in good condition; if you

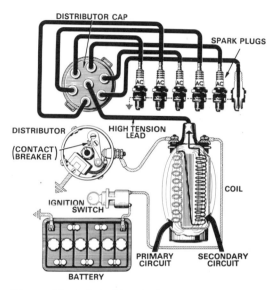

Fig 44. *The basic ignition circuit.*

neglect this chore, the plates will sulphate up and damage will be irreparable. Eventually, the plates of even the best kept battery begin to disintegrate and fail and there is nothing else for it but to buy a new set. I have noticed that when batteries are getting towards the end of their life, they begin to use a lot of water to keep up the electrolyte level. Your boatyard will have a special instrument for testing each of the cells and, if you notice voltage is constantly low, this test should be done. The terminal posts and the cable terminations should always be kept smeared with vaseline petroleum jelly to stop verdigris building up,

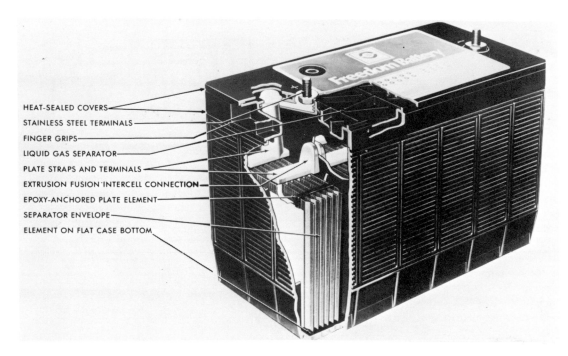

HEAT-SEALED COVERS
STAINLESS STEEL TERMINALS
FINGER GRIPS
LIQUID GAS SEPARATOR
PLATE STRAPS AND TERMINALS
EXTRUSION FUSION INTERCELL CONNECTION
EPOXY-ANCHORED PLATE ELEMENT
SEPARATOR ENVELOPE
ELEMENT ON FLAT CASE BOTTOM

Fig 45. The AC-Delco Freedom battery.

for this will certainly impair the free flow of current as much as a loose connection. Whilst charging, this type of battery will give off explosive hydrogen gas, so don't go checking the electrolyte level with a lighted match.

The tops of the batteries must be kept clean and dry, and it is highly advisable to have a cover over them, arranged so that there is air space beneath it for ventilation and protection against anything dropping from above. The batteries on my own boat are arranged with an acid-proof plastic tray above. This has a one inch lip around and makes a good tray for holding tools when servicing the engines. I am aware that some authorities do not permit batteries to be in the same compartment as the engines, but my boat has full mechanical ventilation, both input and extractor fans being used, and the mast provides natural ventilation above the batteries by means of a specially designed mast head which produces a venturi effect to extract air.

Lead/Acid Battery	Hydrometer Reading					
State of Cell	10°C 50°F	16°C 60°F	21°C 70°F	27°C 80°F	32°C 90°F	38°C 100°F
Fully Charged	1·288	1·284	1·280	1·276	1·272	1·268
Half discharged	1·208	1·204	1·200	1·196	1·192	1·188
Fully discharged	1·118	1·114	1·110	1·106	1·102	1·098

Table 5. Specific gravity readings for a typical lead/acid battery

Make sure that the battery and its box are securely anchored to prevent movement in the event of a knockdown.

A newcomer from the AC-Delco Division of General Motors is the Freedom battery (Fig 45) which needs no maintenance. It has lead-calcium alloys for its grids, and other new developments in its construction to ensure that there is virtually nothing for the owner to do. Terminals of stainless steel are corrosion free, there is no topping up and, perhaps best of all, no charging to be done during the lay-up period; it will retain its charge for up to fifteen months without attention. In some applications the alkaline battery has proved popular because of a similar low maintenance schedule although it does need topping up with distilled water. Initially, they are very expensive compared to the lead/acid types and will probably become a rarity with the new competition from AC-Delco.

The hydrometer is a simple instrument for determining the specific gravity of battery electrolyte, but on no account must a hydrometer from a lead/acid battery be used on an alkaline battery or vice versa. For correct usage, it is important to know the ambient temperature and then take a reading. The table above (for a lead/acid battery) specifies the state of charge at the appropriate temperature and specific gravity reading.

Battery Isolating Switches

These are an important part of any marine installation, as they provide safety from electrical faults and possible fires when the boat is left. They may be small mechanical switches, such as the ones made by Lucas, which fit directly onto the battery terminal post with the heavy duty cable leading from them, or electrically operated solenoid switches whereby low voltage actuates an electromagnet which in turn switches the batteries into full circuit. Solenoid switches must be kept out of damp places and none used except those designed for a marine environment. Servicing at two or three yearly intervals is normally sufficient to keep them in first class condition. A proprietary electrical contact cleaning aerosol followed by a light grade aerosol anti-corrosive lubricant (LPS No 1) is ideal.

Ignition Switches

These may, very occasionally, need similar treatment as above. Although reliable, they can fail, especially when placed in badly exposed positions. Being derived from the automotive industry, the majority are not water or spray proof and must, therefore, be fully protected; check that the wires to the switch are properly secured.

The Coil

This produces the high tension current which provides the spark across the electrodes of the sparking plug. The canister contains two circuits, one the primary or low tension side, fed through the ignition switch from the battery; the other, the high tension side, which provides the spark. The low tension side of the coil is tested by means of a bulb in a bulb holder at the appropriate voltage rating. One wire from the holder is earthed on the engine and the other held in contact with the post normally marked 'CB' or, if not marked, the one that links the coil to the contact breaker. With plugs removed so that the engine can be turned easily by hand, the bulb should soon indicate by regular flashing if the primary circuit is working; take off the distributor cap and note how, when the contact breaker points are open, the light shines and how it is doused when the contacts close.

The high tension windings are then tested, initially by laying one of the plugs on the cylinder head so that its body is fully earthed and turning the engine over. It is highly dangerous to do such a test if there is any petrol vapour about which can catch fire or explode; make sure everywhere is extremely well ventilated first. A far safer way is by means of a Smiths Industries high tension tester, which is inserted onto the end of the spark plug with a high tension lead from the distributor head to the plug; turning over the engine will cause the tester to light up. If the coil is producing the spark but there is nothing when the high tension is passed through the distributor, the fault will be in the distributor or the high tension leads. If there is no spark at the plug or glow from the tester, the high tension lead should be taken off the head of the distributor cap. Holding the lead about a quarter inch from the engine, the test is repeated. A good fat blue spark should jump to earth from the

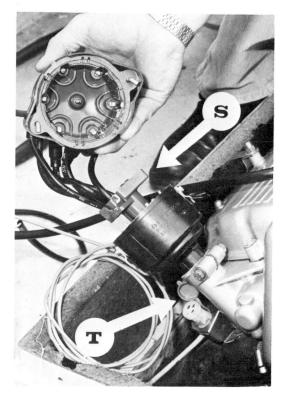

Fig 46. The Distributor on the 6-cyl BMW.
T) Clamping bolt that must always be kept tight unless major timing by specialist is undertaken. S) The Distributor Rotor; this can be removed to immobilise a craft when it is left on moorings.
The distributor cap is in the mechanic's hand showing the central HT electrode which distributes, via the rotor, the HT current to the six outer contact points.

lead, so make sure you are holding the wire by some insulated means or it will give you a nasty shock.

The Contact Breaker and Distributor

Both are parts of one typical unit shown in Fig 46, which is the BMW 6 cylinder engine with its overhead camshaft driving it through a right angle drive gear. The unit is inserted into the main casting housing and secured by means of a clamping bolt (T) just above the arrow. You should *always* put a scratch mark across the distributor casting to the engine casting to ensure that the unit is replaced in precisely the same way. The basic engine timing will then be undisturbed. It is important that the clamp is not disturbed by anyone who does not fully understand the importance of correct timing.

The high tension spark must be sent to each plug at precisely the correct time if it is to initiate the fuel burn to produce power. Rotating the whole body of the contact breaker drastically alters this, and the engine manufacturer specifies the correct timing in relation to TDC (top dead centre). The timing marks are found under a small plate on the engine flywheel while a metal arrow or mark is fixed on the main casting. These marks should be cleaned up and a little white paint rubbed into them so that they show up. An ignition timing light, such as that made by Smiths Industries, is connected by removing the high tension lead from No 1 cylinder spark plug and connecting it to the male lead on the light, while the female connection is placed on the spark plug terminal screw. The No 1 cylinder is the furthest forward from the gearbox on in-line engines, or the front left cylinder in V configurations made by GM, and front right on Ford; in any case, the engine manual should indicate the firing order. Even when you are going to do a simple job like clean the plugs, it is essential for you to know this and be able to replace high tension leads on the correct plugs. For the unsure, a dab of paint on the No 1 HT socket on the distributor and temporary masking paper tags on each of the leads, correctly numbered, will ensure you get things back the way they were. On small four cylinder engines, the length of the HT leads is often a quite sufficient guide, as they will not stretch to fit the wrong spark plug; many a panic has been caused by fitting the wrong lead and getting very rough running as a result. Before timing can be checked, two small operations have to be completed. First, the automatic vacuum advance device has to be disconnected from the manifold to the distributor and, secondly, the hole at the manifold must be plugged to ensure that the normal fuel mixture is being supplied to the engine.

Automatic devices for advancing and retarding the spark greatly assist smooth running from start, during warm up and for the full range of speed. These devices are able to move the base plate on which is mounted the contact breaker, and two kinds are often used: a mechanical one based on the centrifugal force produced by rotation, located on the lower part of the contact breaker body; and a supplementary vacuum device driven from the atmospheric depression at the carburettor venturi. By changing the relative position of the points in relation to the central cam (which opens and closes them) a degree of advance or retard is achieved.

With all connections properly made and a warm engine, the timing light is pointed at the timing mark which should appear stationary. The degree of advance or retard will be apparent by its position in relation to the reference mark on the clutch housing or crankcase. If the timing is out of adjustment, the clamping nut is undone and rotated in the direction of the rotor arm to advance ignition, or vice versa to retard it; this direction is often marked on the rotor itself. Some Lucas distributors have a vernier scale of advance and retard incorporated immediately near the vacuum bellows. Small adjustments of advance or retard can be achieved by moving the knurled ring in the direction of 'A' (advance) or 'R' (retard) marked on the casting.

To check the automatic advance/retard mechanisms is quite simple. With the vacuum one disconnected, accelerate the engine and note the advance produced by the mechanical centrifugal mechanism. Check with

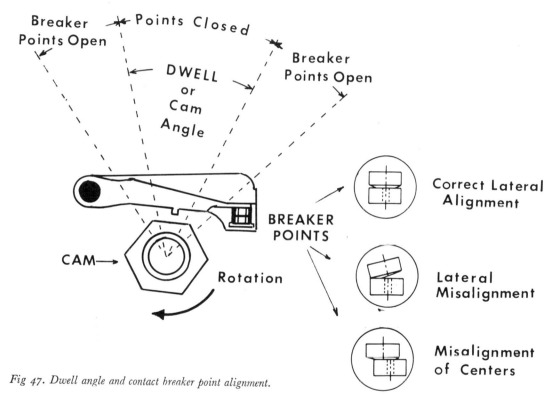

Fig 47. Dwell angle and contact breaker point alignment.

the vacuum pipe reconnected. If this is satisfactory, stop the engine and, as a crude but simple test, by means of a small plastic pipe connected to the manifold link pipe apply suction and pressure by blowing. When this is done, observe the contact breaker mounting plate which should move slightly back and forth with the suction and pressure; very occasionally, the diaphragm in the vacuum advance mechanism ruptures and has to be replaced. The vacuum device can now be more fully tested by reconnecting the vacuum line (making sure the plug is removed) and again running the engine. As it is accelerated, the timing light shining on the marks should show a greater degree of advance compared with when it was undertaken by the mechanical device alone. Although the amateur can check the basic functions of the timing, I would stress that this is really the province of the expert. Timing is closely linked to carburettor tuning and it will damage the engine if these items are not synchronised correctly. Another inhibiting factor is that manufacturers of modern power units specify the dwell angle, (Fig 47), the angle through which the points remain closed, as opposed to the angle through which they open, and this check requires a dwell meter – an expensive electronic instrument.

Breaker Points, Condenser and Rotor

The breaker points are small, inexpensive items that do a great deal of hard work and need regular servicing and replacement. The condenser minimises arcing across the points, but eventually they will burn out; small pitting occurs in their surface so that correct electrical integrity is lost. The breaker points are found under the distributor rotor which is easily slipped off its drive shaft (Fig 46S). The distributor rotor should be checked for cracks and to see that the contact on the centre and running to the outer tip is bright and clean. A spare should be carried and it is worth removing this item when you want to immobilise the boat for security reasons. It can be cleaned with one of the proprietary electrical contact cleaning aerosol sprays, such as those made by LPS.

Badly burned contact points are sometimes the result of installing a radio condenser on the distributor side of the coil, or an incorrectly set voltage regulator. More often, though, there will be merely pitting or a slight transfer of metal from one point to the other. Running the engine at other than normal speeds, poor contact point alignment or a failing condenser could be the culprits. Fig 47 shows the way points should be aligned and the result of incorrect alignment. Occasionally, you can achieve a small saving by smoothing off the old points with a fine magneto file, but this must be done very carefully. They must be perfectly flat and in perfect alignment; a badly filed set will last no time at all. A new set of points is relatively easy to install and once they are in place all you need to do is check the gap. To do this, rotate the engine until the small rubbing block on the moving point comes onto the top of the cam. With the other contact point securing screw fairly loose, insert a feeler gauge of the correct gap setting between the points. Checking that they do not move in gap or alignment, secure the screw to the fixed point. The cam that drives the points through the rubbing block should be lubricated with a proper distributor cam oil – not engine oil. A little light oil should also be applied to the rotor bearing surface of its drive shaft. Old points are often not worth resurfacing, as it is the spring tension that is

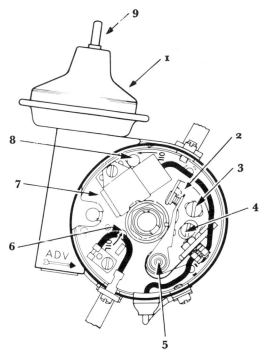

Fig 48a. Adjusting the contact breaker points.
The two fixing screws holding the fixed point are slackened off and the engine turned until the moving point is in its fully open position on the top of the cam. The fixed point can then be moved towards the other point, the correct distance being achieved with a feeler gauge. The screwdriver is used to tighten down the fixing screws. Check the gap again, and see that the two points are perfectly aligned.

Fig 48b. Servicing points in the lower contact breaker assembly.
1) Vacuum diaphragm housing that controls the automatic ignition advance from the depression in the carburettor manifold.
2) Contact breaker points. 3) and 4) Fixed contact adjusting screws. 5) Moving contact pivot post needs light lubrication.
6) and 8) Light oiling points for the centrifugal advance mechanism which is housed in the lower part of the unit.
7) Condenser. Suspect failure if there is bad pitting on the points or if part of one point is transferred to the other as a little hump of metal. The condenser needs a professional to test it properly. 9) Vacuum pipe to the carburettor manifold.

causing trouble in the engine; correct tension is essential to decent contact life. Too great a tension leads to rapid wear of the breaker arm and rubbing block, causing the breaker point gap to close up and altering the dwell angle. If the spring tension is too weak, the breaker arm will flutter at high engine rpm, resulting in misfiring. After replacing points, it is always advisable to check the timing.

The Distributor

This is easily recognised as a black plastic cap that has to be removed to get at the points; while it is off, check that it is in good condition. Fig 46 shows the distributor cap in the mechanic's hand. Test it by checking that the central sprung contact is free and clean so that it will bear on the rotor arm properly. The cap should always be kept immaculately clean, as it handles the high voltage to the plugs. Dust holds moisture and this in turn can cause a leakage of high voltage current to earth. It is not uncommon, as Fig 49 shows, for the head to crack and, though this is very difficult to see, it will certainly cause the unit to misfunction. The place to look is on the inside of the casing where the channels carrying the high tension leads from the coil to the central electrode and the spark plugs run. A crack between two of these channels, though only a hairline, will be a source of power loss, if not complete failure. Check that the pointed grub screws which secure the

Fig 49. A cracked distributor cap; the hairline crack is sufficient to make the cap completely useless as it allows the HT current for the spark to escape to earth.

high tension leads are tight; the leads themselves sometimes get damaged and certainly age. A crack in a cable conductor does not always show up on the outside, so it is sensible to renew cables fairly regularly if they show signs of ageing. The cap can be cleaned with a suitable aerosol and dried with a lint-free cloth before replacing.

The Spark Plugs

If there was one part of the engine I would

GAP BRIDGED	**OIL FOULED**	**CARBON FOULED**
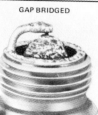		
IDENTIFIED BY DEPOSIT BUILD-UP CLOSING GAP BETWEEN ELECTRODES. CAUSED BY OIL OR CARBON FOULING. IF DEPOSITS ARE NOT EXCESSIVE, THE PLUG CAN BE CLEANED.	IDENTIFIED BY WET BLACK DEPOSITS ON THE INSULATOR SHELL BORE ELECTRODES CAUSED BY EXCESSIVE OIL ENTERING COMBUSTION CHAMBER THROUGH WORN RINGS AND PISTONS, EXCESSIVE CLEARANCE BETWEEN VALVE GUIDES AND STEMS, OR WORN OR LOOSE BEARINGS. CAN BE CLEANED IF ENGINE IS NOT REPAIRED, USE A HOTTER PLUG.	IDENTIFIED BY BLACK, DRY FLUFFY CARBON DEPOSITS ON INSULATOR TIPS, EXPOSED SHELL SURFACES AND ELECTRODES. CAUSED BY TOO COLD A PLUG, WEAK IGNITION, DIRTY AIR CLEANER, DEFECTIVE FUEL PUMP, TOO RICH A FUEL MIXTURE, IMPROPERLY OPERATING HEAT RISER OR EXCESSIVE IDLING. CAN BE CLEANED.
WORN	**NORMAL**	**LEAD FOULED**

IDENTIFIED BY SEVERELY ERODED OR WORN ELECTRODES. CAUSED BY NORMAL WEAR, SHOULD BE REPLACED	IDENTIFIED BY LIGHT TAN OR GRAY DEPOSITS ON THE FIRING TIP. CAN BE CLEANED.	IDENTIFIED BY DARK GRAY, BLACK, YELLOW OR TAN DEPOSITS OR A FUSED GLAZED COATING ON THE INSULATOR TIP. CAUSED BY HIGHLY LEADED GASOLINE. CAN BE CLEANED.
PRE-IGNITION	**OVERHEATING**	**FUSED SPOT DEPOSIT**

IDENTIFIED BY MELTED ELECTRODES AND POSSIBLY BLISTERED INSULATOR. METALLIC DEPOSITS ON INSULATOR INDICATE ENGINE DAMAGE. CAUSED BY WRONG TYPE OF FUEL, INCORRECT IGNITION TIMING OR ADVANCE, TOO HOT A PLUG, BURNT VALVES OR ENGINE OVERHEATING. REPLACE THE PLUG.	IDENTIFIED BY A WHITE OR LIGHT GRAY INSULATOR WITH SMALL BLACK OR GRAY BROWN SPOTS AND WITH BLUISH-BURNT APPEARANCE OF ELECTRODES, CAUSED BY ENGINE OVERHEATING. WRONG TYPE OF FUEL, LOOSE SPARK PLUGS, TOO HOT A PLUG, LOW FUEL PUMP PRESSURE OR INCORRECT IGNITION TIMING. REPLACE THE PLUG.	IDENTIFIED BY MELTED OR SPOTTY DEPOSITS RESEMBLING BUBBLES OR BLISTERS. CAUSED BY SUDDEN ACCELERATION. CAN BE CLEANED.

Fig 50. Checking spark plug condition.

recommend the amateur mechanic to start his training on, it is looking after the spark plugs. There is little to go wrong provided a proper plug wrench is used and the plugs are not overtightened when they are replaced. The condition of the plug tells the story of the engine. Study Fig 50.

Gap bridging is most commonly found on the old two-stroke engines which had a nasty habit of shorting out just as one approached the mooring or harbour wall. The modern two-stroke oils did much to get rid of this menace, but in four-stroke engines such fouling may be associated with excessive oil in the combustion space. When you are buying a second hand four-stroke and discover oil fouling, expect that the engine will soon require serious mechanical work. You can delay this a little by putting in a hotter plug to burn off the oil; Fig 51 explains the terminology for a hot or cold plug. You will notice that the length of the ceramic insulating core through which the central electrode runs, differs on the hot and cold plug. On the hot plug, the path for the heat to travel from the electrode tip through the body of the plug to the engine's casting is longer, therefore the plug runs relatively hot. On the cold plug the shorter central electrode dissipates the heat faster. Carbon fouling is often the result of too much idling, such as when trolling for mackerel. Running the engine at an idle keeps it cool, allowing soot to build up. If you haven't been running the engine at an idle, look carefully at the other causes of

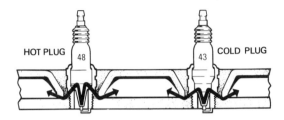

Fig 51. Hot and Cold spark plug.

soot – especially that over-rich mixture which wastes fuel and money. Old, worn plugs can be cleaned and used in winter to block the plug holes while good plugs are cleaned for future use.

Pre-ignition is a most dangerous condition as it will result in replacing plugs, pistons and bores – all highly expensive. Any metal deposited on the plug would be bits of piston; overheating should already have been indicated by the engine's temperature gauge and these deposits on the plug act as confirmation. A cooler plug can be put in as a last resort, but it is far more important to find the cause of overheating which, if ignored, leads on to pre-ignition.

Before removing plugs from the engine, clean round the seating, to prevent dirt from dropping into the cylinder. As you remove them (one at a time if you are worried about getting the leads mixed up) see that the buttress top insulator (Fig 52) is not cracked; it is easily done if the plug spanner slips. A crack or dirt on the buttress will allow HT current to leak to earth down the outside

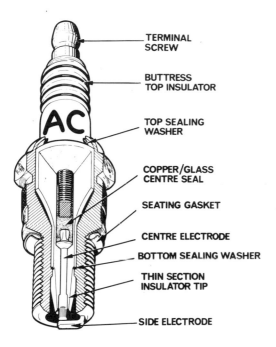

Fig 52. Spark plug terminology.

- TERMINAL SCREW
- BUTTRESS TOP INSULATOR
- TOP SEALING WASHER
- COPPER/GLASS CENTRE SEAL
- SEATING GASKET
- CENTRE ELECTRODE
- BOTTOM SEALING WASHER
- THIN SECTION INSULATOR TIP
- SIDE ELECTRODE

Fig 52. Spark plug terminology.

rather than arc across the plug points inside. Look at the seating gasket to see that the plug has been properly seated on the cylinder head. Discoloured or corroded surfaces indicate that there has been gas leakage because of insufficient tightening or uneven seating. If the gasket is entirely flattened it is a sure sign that the plug has been over-tightened. Plugs can be taken to a garage for grit and air blasting.

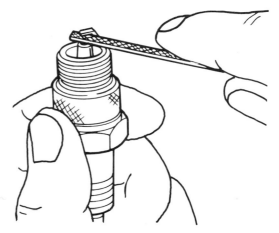

Fig 53. Dressing spark plug electrodes.

To avoid plug trouble, always carry a spare set. Self-servicing should be a last resort, but you can brush soft deposits off with petrol and then scrape off the harder deposits with a knife blade. In doing so, you will probably damage the insulator, leave behind metal scrapings in the plug body to cause a short, and you will not reach right up the plug to clean it completely. If a plug sparks well in the open, it is no guarantee that it will spark when under compression in the cylinder. After cleaning, the electrodes should be dressed as in Fig 53 with a small magneto file to get both centre and side electrodes parallel. The gap between them is adjusted to the engine maker's specification using a feeler gauge. Only bend the side or earth electrode to do it. Never touch the centre electrode as this will crack the ceramic insulation and ruin it.

When refitting the plugs, the ideal is to use new gaskets which will ensure the correct heat transfer between plug and cylinder head casting and make a gas-tight seal; but I recognise that this almost never gets done. Turn the plug down until it is finger tight on the gasket, then give it one half turn with the plug spanner. Table 6 (below) gives the proper tightening torques, should you have a torque wrench to hand.

| Plug Thread | Correct Tightening Torques | |
	Cast Iron Heads	Aluminium Heads
10mm*	12 lb/ft	10 lb/ft
14mm*	25 lb/ft	22 lb/ft
18mm*	30 lb/ft	25 lb/ft
7/8ins	35 lb/ft	30 lb/ft
18mm taper	15 lb/ft	12 lb/ft

*Available in Heli-Coil repair kits (12mm also made).

Table 6. Spark plug tightening torques

Plugs screwed into light alloy heads need a deal less torque if you are not to damage the casting. It is very easy to cause such damage and, if you are unfortunate enough to do this, I would recommend the Heli-Coil spark plug thread repair sets. (Fig 12). It may be uneconomical to purchase a full kit, especially if a local garage stocks the Heli-Coil system, but a kit is a great deal less expensive than a new head. While you have the head off doing one plug hole, you may as well do the lot. Check that you order the correct reach as the basic sizes are also available in various reaches.

The Starter Motor

Although, basically, the starter motor is only an electrical motor driven from the batteries, it is a complex piece of equipment and servicing is beyond the scope of the amateur mechanic. The starter pinion must engage gently with the flywheel before full electrical power is applied, or the gear teeth will be stripped off. Accidentally pressing the starter when the engine is running is damaging as is trying to start before the engine has come properly to rest after an initial attempt.

Be gentle with the starter button; don't engage it for a long time if the engine doesn't fire up.

Fig 54 shows types of damage resulting from normal wear and abuse to the starter pinion. It is within the amateur's scope to remove the electrical leads from the starter motor after switching off the batteries, to take the starter motor off the engine and check the pinions to see that all is well. If there are signs of serious wear, the unit is best taken to an official service station as the inside, as well as the pinion, will probably need attention.

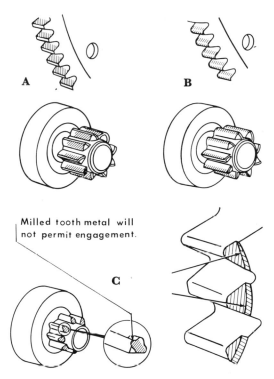

Milled tooth metal will not permit engagement.

Fig 54. Checking starter motor drive pinions. A) normal wear. B) small wear pattern. C) milled condition showing excessive wear on two or three teeth.

After the pinion engages, partial switching of the power allows a slow rotation and, when fully engaged, full power from the battery is switched in to turn the flywheel over smartly, so creating the first compression and firing of the fuel in the cylinders. A weak battery will not supply its full voltage to the starter motor to do this, and you will get poor starting, if any at all.

Inside the starter motor there are all sorts of springs and sliding parts which engage and disengage, allowing partial and full switching of current. The most likely causes of trouble are weak springs or dirt on the shafts and sliding components which stop the correct sequence of movement.

When taking the starter motor off the engine for servicing clean the engine flange to which it is bolted, and cover the opening with a lint-free cloth. Most starter motors are run until they begin to give trouble, and this is wrong because you may be out at sea when it happens. With the low engine hours achieved by most boats, a full maker's service is advisable long before you would normally expect it. Failure is more likely to be lack of use and standing idle rather than wear and tear. I would suggest a full service after five seasons.

The most frightening thing that can happen to the starter motor is for it to remain engaged when the engine starts. The fly-wheel is now driving the pinion on the starter motor to phenomenal revolutions which will quickly cause serious damage. Stop the engine immediately if this happens.

In general, manufacturers engineer their products to ensure that no damage occurs when the engine rocks back or fails to disengage. With Prestolite equipment on Mercury engines, for example, if the engine starts but does not reach a high enough speed to disengage the detent pin, a ratchet type clutch on a screw shaft lets the pinion overrun the armature shaft. When this occurs, a light buzzing sound is heard; this is caused by the clutch ratchetting. Accelerating the engine releases the clutch and the sound will disappear. Lucas fit an overspeed device on some of their units, which is designed to protect the armature bearings and brushes against over-rapid acceleration. The rapidly rising current from the dynamo operates a relay which cuts the electrical supply from the battery to the starter motor. A problem can occur when there is a faulty relay or dynamo which may prejudice starting. If this happens, the starter motor can be made to function in an emergency by temporarily bridging the two main terminals on the relay type ST, so closing the starter solenoid circuit and cutting out the overspeed protection.

Starter motors are, unfortunately, usually located low down on the engine. Make sure you cover them, when it comes to the winterisation programme, especially if there is a drain tap for the cooling system just above them. They can be easily sealed off with a plastic bag and masking tape if they are not to be removed for winter servicing.

Dynastart Systems. On small marine engines a Dynastart is often used. This unit is a combination of a dynamo, which generates sufficient power to keep a small battery charged, and a starter; the unit is usually driven from a belt drive taken from the flywheel. This belt can be the cause of problems if it is not kept in good condition and at the correct tension; any excess slackness makes both starting and generating suffer—aim for half an inch deflection of the belt at maximum. Always carry spare belts.

Early SIBA Dynastarts were of quite low generating power but in recent years, with the growth of the use of electrical power on boats, the units have been uprated to give quite generous output for their relatively compact size. For example, the SIBA DS418 has an output of 60 watts and a starter motor power of 0·8hp, while the Bosch starter/generator LA/Ej 90/2000 has a 135 (90 continuous) generating capacity with 1hp starting. This latter unit enables Volvo Penta to recommend that the battery capacity could be increased from 32ah capacity to 60ah with the Bosch unit.

Apart from keeping the unit clean and dry, there is no work that the amateur should do on these units. They should be taken to an agent for servicing.

Magnetos are again only used on small petrol engines and their use has, over the years, declined in popularity. This is because the modern yachtsman, even on the smallest boat, wants push-button electrical power for services as well as ignition. I still favour marine engines that offer the security of a second starting system. Magnetos provide a spark with great simplicity although, if the truth were admitted, on inboard engines in marine conditions they can be a trial to a saint.

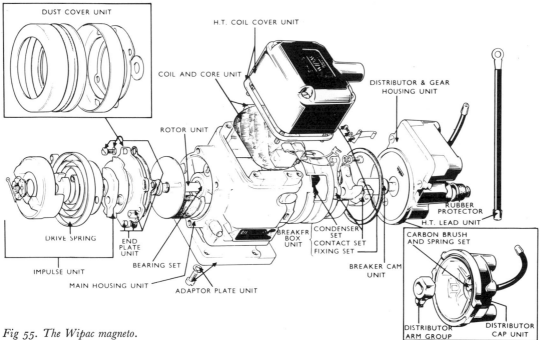

DUST COVER UNIT

H.T. COIL COVER UNIT

COIL AND CORE UNIT

DISTRIBUTOR & GEAR
HOUSING UNIT

ROTOR UNIT

RUBBER
PROTECTOR

H.T. LEAD UNIT

DRIVE SPRING

END
PLATE
UNIT

BREAKER
BOX
UNIT

CONDENSER
SET

CONTACT SET

FIXING SET

CARBON BRUSH
AND SPRING SET

IMPULSE UNIT

BEARING SET

BREAKER CAM
UNIT

MAIN HOUSING UNIT

ADAPTOR PLATE UNIT

DISTRIBUTOR
ARM GROUP

DISTRIBUTOR
CAP UNIT

Fig 55. The Wipac magneto.

Although magnetos are tending to become museum pieces, it is worth noting that the WIPAC group in Buckingham, UK, offer a service to magneto users and are keen to be of assistance in servicing any units they made in the past.

Fig 55 shows one of the WIPAC A series magnetos which were found on a number of marine engines, including Stuart Turner Coventry Victor and Britt. Lucas also made the SR series of magnetos but basically they performed the same functions and their servicing, though varying in details, still has a number of basic points. Make sure the magneto is kept thoroughly dry. Easier said than done in a marine environment, but certainly it should be removed from the boat in winter and never wrapped in plastic so that condensation is trapped. In removing the magneto, remember it is, like the distributor, a timed unit that must be replaced in exactly the same position as it was when the engine was working. Look in the maker's literature to see if it has timing marks; if not, make your own. Servicing consists of seeing to points, lubrication and cleaning of the impulse starter, when fitted. Magneto points do exactly the same job as the conventional ones on coil ignition engines. There is a fixed and a moving point, and it is important to gap them correctly; the engine manufacturer recommends the gap, but WIPAC units were usually set at 0·015in and Lucas ST at 0·010in to 0·012in. You can still buy magneto files which are used for cleaning up and making true the points, as in Fig 47, but a fine carborundum stone, silicon carbide paper or even emery cloth will do the job, just so long as the debris is thoroughly cleaned off with a spot of surgical spirits. Although some magnetos had platinum contacts which were good for long service, other types formed a film of hydrated tungsten oxide over the winter months or when little used; this must be cleaned off as already described.

The impulse starter will be found on some magnetos. At the instant of firing, the rotor is accelerated at high speed to produce a greater intensity of spark and, at the same time, firing is delayed until the piston is approximately top dead centre, i.e. retarded. This device ensures that the engine will fire up even when it is gently turned over compression rather than swung madly with the possibility of a snatch as the piston comes over top dead centre. Incidentally, always keep the thumb on the same side of the starting handle as the fingers, or you might break it if the engine backfires, snatching the handle from your hand. The impulse starter unit is usually fitted under cover which should be removed for access. It may be clogged with dust and dirt so that the trip arm fails to engage or disengage or the unit is sluggish in action. With the utmost caution, and taking care that no paraffin works its way into the magneto housing, the unit can be flushed out to remove the dirt. An electrics cleaning aerosol is perhaps a safer, though a more expensive way of cleaning. The unit should be lubricated with a thin machine or engine oil – never grease – and the cover replaced.

Magneto lubrication points differ slightly with makes. The Wico type A models have two spring oilers, one of which should be filled to overflowing with a light grade engine oil, such as Castrolite, every 200 hours or each boating season. Every couple of seasons the cam grease pad needs attention. The pad is removed and a light grade of motor transmission grease worked into it. The Lucas SR magnetos should have a spot of clean engine oil applied to the contact breaker pivot post once each season. On some models which have a gear drive to the rotating electrode, a few drops of light machine oil should be applied through the distributor gear oil hole. Be sparing with it, for trouble can be caused if a drop gets onto the points. The earlier SR models have felt pad lubrication of the cam similar to the Wico already described, but later models have a porous sintered iron cam which needs to be re-impregnated with a light grade of engine oil. The sintering process of manufacture ensures a porous grade of iron which will retain oil and is therefore self-lubricating for a period of time.

Automatic timing control will be found on some magnetos although in the SR series this is usually found separate from the unit, on the engine timing case. It works on the principle of centrifugal force advancing the spark as the engine starts and, as speeds fall and the centrifugal weights close, the spark is retarded. Normally, no maintenance is needed for long periods, but every couple of years, weights, springs and toggles should be examined and lubricated with a medium viscosity engine oil. As I mentioned earlier, the magneto is a timed unit so don't fiddle about with it unless you understand it. The engine manufacturer will supply correct information but in an emergency Lucas recommend the following steps as a temporary measure. Magnetos fitted with automatic timing control: contact breaker points to open when the cylinder is under compression at top dead centre (TDC). Magnetos with or without impulse starters: contact breaker to open when the piston in the cylinder under compression is $\frac{5}{16}$in to $\frac{3}{8}$in (8–9·5mm) before TDC. Some magnetos have an earthing switch on the side; see that it is kept clean and working correctly.

One final point – check the HT leads at least once a season, and test them as advised for the distributor.

Fig 56. Correct generator drive belt tension should be about right when there is a half-inch deflection on the longest run.

The Electrical Generating System. In ignition systems such as the conventional coil-based system (Fig 44) and the modern capacitor discharge (CD) system, which I will deal with later, a battery fed from a generator supplies the electrical power needed on the majority of inboard power units. Outboard engines and some small saildrive units use the modern CD system, but with a magneto. However, what concerns us here is how to keep the battery fully charged. On older engines – and there are a great many about – the dynamo does this job, while on more modern engines the alternator has replaced the dynamo. Both types of machine are driven from the engine by belt drive so the first job is to ensure that this belt is always kept to the proper tension. Too little tension will lead to slipping and belt wear. Too much tension and the machine's bearings will be damaged. Fig 56 shows the front of the BMW marine engine, where the toothed belt drives the water pump and the alternator from the main crankshaft pulley. On any marine engine, it is suggested that a half inch deflection on the longest run of the belt drive will give the correct tension. Sometimes toothed rubber belts are employed, and these are tightened or loosened by removing or replacing shims under the drive wheel, which is made of two halves fastened together with four bolts; adding shims reduces the tension and taking them away increases it. The BMW units have this method for their raw water pump drive. On systems using a plain belt, the tension is adjusted by loosening off three bolts which hold the machine in place, and moving it down on the adjusting slot to get the correct tension. Make sure that you do not damage other parts of the engine, especi-

ally piping and wires. It is a good idea to get a short length of timber and insert it between the engine casting and the alternator or dynamo to get a leverage on it. I suggest that the alternator should be removed from the engine and taken for a manufacturer's service every four or five years; label the electrical leads before removing. Problems arise because of lack of use rather than too much.

When parts of the electrical system are being serviced or disconnected, see that there is no chance of short circuits occurring due to wires touching or accidental starting when batteries are disconnected from alternators. The diodes and transistors in alternators are destroyed by reversing polarity or by trying to run them unconnected to the batteries.

Never 'flash' connections to the alternator to see if there is current flowing; no matter how briefly you do this, transistors may be ruined. Incidentally, a part of Lucas alternator equipment is a fast fuse. This consists of a metal and moulded plastic box containing a special fuse made with high accuracy, the rating being specially chosen to protect the alternator rectifier diodes from damage caused by reverse polarity battery connections. If you take the precautions already suggested, you will not blow this fuse, but if for some reason you do or there seems to be a fault in the system, check out all connections and polarity *before* using the replacement fuse which is carried in the fuse box. They are quite expensive little items and it is no use blowing the replacement if you have not found the fault that caused the first to go.

The Dynamo. This should be removed from the engine every three years, even if it appears to be working perfectly. A professional service is needed because the carbon dust gathers in the area of the carbon brushes and the commutator, causing the brushes to wear away slowly and need replacement. The dust must be cleaned out and the commutator inspected for wear; the springs that keep the brushes in contact with the commutator lose their tension and poor contact is made. On some old dynamos, you may need to oil the end bearing away from the commutator; be sparing with oil as you do not want to get it onto the windings. The voltage output from the dynamo is controlled by a regulator, which is another black box full of mystery and, unless it is thoroughly understood, should not be tampered with; its job is to keep the voltage output of the dynamo at a prescribed level. There are contacts on the regulator which the engineer will adjust with the aid of proper tools and a meter. The regulator is usually of the vibrating armature type, which has to be kept in proper adjustment, and this is perhaps another factor which has favoured the alternator, as these usually have self-limiting voltage control or a solid state voltage regulator which needs no attention. Both dynamos and alternators are affected by heat and should always be kept well ventilated.

Electronic Capacitor Discharge Ignition Systems. Knowing when to leave parts of the engine strictly alone is probably as important as knowing how to service others. Certainly, when it comes to electronic ignition systems, nobody but the trained mechanic should tamper with them. The reasons are two-fold. First, they usually have a means of storing electricity which will eventually be released to the spark plugs. This is stored in a capacitor, and if you do not know what you are doing, you stand a good chance of discharging the HT circuit into yourself, with shocking results. Secondly, like the alternator, the circuits employed have a host of diodes and transistors which are polarity conscious and will fail if flashed. Your engine manual should indicate if there are oiling holes, but these are the only points you will find as there are none to adjust, such as the ones on the conventional ignition.

With the Seagull saildrive, a breakerless magneto energises the CD system, the energy required for ignition being stored in a capacitor until the desired time of ignition.

When the magnet mounted in the flywheel is rotated past the coil core on the stator plate at sufficient speed, a voltage is generated in a feed and trigger winding. The voltage in the feed winding is used to charge the storage capacitor. This charge remains in the capacitor until a trigger voltage activates a semiconductor switch to release the energy into the primary winding of an ignition coil. The resulting voltage produced in the primary is transformed up to the ignition voltage at the spark plug, firing the engine and repeating the process.

Incorporated in the WIPAC design are features which produce oscillations between the capacitor and ignition coil primary to extend spark duration. Also, from starting to around 1500 rpm, automatic advance in ignition timing is achieved. Spark rise time is fast, so that fouled sparking plugs continue to function for much longer.

Take care when testing the spark at the plug that it is properly earthed onto the cylinder, or that the ignition switch is off when the engine is rotated. On some capacitor discharge systems damage could result if this is not observed. Check manufacturer's instructions.

Although at this stage of the development of electronic ignition systems, it is somewhat frustrating for the owner to be limited in the work he can do, with computer control of marine petrol engines now well established, we shall all have to get used to the idea. In time, perhaps, the unit will tell us on a display what is wrong with it. Certainly, the next generation of petrol marine engines will be able to be plugged into a shore based computer to give a complete print out of the state of the engine and which parts are not functioning correctly. Still, I don't suppose there will ever be an engine that lays itself up and gets itself ready in the spring for another boating season.

seven

Gear Boxes and Outdrives

If you are not mechanically minded, you will be relieved to know that gear boxes and outdrives are beyond simple skill when it comes to taking them apart. However, there is much you can do to prolong their life and look after them. If you give them a regular look over and see to the recommended servicing, you will ensure that expensive noises are minimised.

Mechanically Linked Gear Changes

Small launch engines with a mechanical box may still be found. Forward, neutral or reverse gears are engaged by means of a lever; these boxes need too much mechanical force to be moved by the single lever remote controls on modern boats. In the past, mechanical boxes were sometimes linked and shifted by means of complex bell cranks and connecting rods, which needed a great deal of energy. If your boat has such a system, the bell cranks and links should be checked and greased for security and efficiency. Cheaper systems have forked links on the ends of rods secured to bell cranks with a drop through pin. Better quality units have ball joints on the cranks and rods with sockets that can be tightened and split pins locked to give a smooth action without backlash. Check that balls and pins are not fatigued: distortion of any metal fitting is a sure sign that it has been over-strained at some time and a little extra strain may finish it off; socket joints can be

filled with an appropriate grease when re-assembling. In older boat installations you may well find the throttle control linked to the steering position by a similar system which will need servicing in the same way.

Mechanical Gear Boxes

In recent years, the German Hurth gear box has made a considerable impact on the market because it can be used with cable control. Its beauty is its compact size, as well as the advantage of a purely mechanical box with little to go wrong except the friction drive plates; you have to get used to the mechanical click as the gear is changed. Gears should be treated carefully: don't, for instance, crash them from forward to reverse at high speed; keep them topped up with the correct grade of oil. If you are rough with them, the excess strain imposed on engine and gear box will have to be paid for at some stage.

Mechanical boxes depend on brake bands or friction bands which are applied to gear trains, so that the engine drive shaft can drive different sets of gear trains for the forward and reverse motion at the propeller shaft. Every ounce of excess friction, needlessly applied, ensures that these parts wear out or need adjustment and the boat will begin to lose drive. You may well not notice this and, gradually, friction will wear parts away even faster until serious repairs or adjustments are necessary. Some mechanical boxes come with simple instructions for adjusting the brake bands. A cover is usually found directly above the bands in the gear box casing. Following the maker's instructions for adjustment is quite simple, and generally only a spanner and screwdriver are needed.

The objects most likely to damage gear boxes are ropes or plastic debris wrapped round the propeller. The friction plates are wrecked when the engine tries to drive the stationary shaft. When this happens, the drive should be immediately disengaged and the engine stopped until the obstruction is cleared. This applies also to other types of box, since shafting and brackets can be seriously damaged. You may have to get over the side and clear a tightly wound obstruction with a sharp knife or small saw. If a rope or line is the culprit and the end can be caught up with a boat hook, it may be possible to heave on this while another crew member winds the shaft round from the engine compartment in the opposite direction to normal rotation, with the unit out of gear.

Reduction Gears

These are sometimes added to the output end of the gear box, mechanical and hydraulic types, to reduce the final propeller shaft r.p.m. Modern engines usually derive their power by producing modest torque multiplied by high speed. Ancient marine engines had very low r.p.m. but enormous torque to produce their power. Although a little power is lost every time it is put through a gear, reduction gears increase efficiency by increasing the torque in order to swing larger diameter propellers. The speed a propeller needs to turn is, of course, dependent on the hull shape. Displacement craft, such as fishing boats, swing the biggest diameter propeller possible to gain the highest drive efficiency for that type of hull. High speed planing craft often have propellers with a straight through drive with no reduction. There is little to do to service reduction boxes, topping up with the correct grade oil being about all they need. However, be prepared for problems when it comes to emptying these boxes. Some drain plugs, as on TMP 12,000 boxes, will be situated on the underside of the box casting. Because of the angle of installation, you usually find that the drain plug is situated down between the two sides of the modern fibreglass hull moulding that form the keel section. Getting a spanner into such a confined space is difficult at best and, on some sailing boats, well nigh impossible. I am delighted that some of the more enlightened gear box manufacturers are now angling the final output drive so that the engine can actually be installed at a more sensible level. This not only gives better access but ensures that engines can be put under wheelhouse floors more comfortably. Mechanically speaking, the low installation angle can be much kinder to the engine's lubrication system in cases where high angles of trim come perilously close to the maximum allowable.

Although engine builders often provide a sump pump for the main engine oil and, occasionally, incorporate a two way cock so that the same pump would empty the gear box oil, the facilities provided on many marine engines are far from perfect. The situation can be retrieved by use of one of the modern pumps.

A reduction box does not have its own dipstick, but the level is usually determined by a removable threaded plug on the side of the casting. On filling, the instructions will usually say fill until oil comes out of the plug then replace it. With some thick oils, it is quite easy to overfill as high viscosity means that there is some delay between the last oil being put in and the oil finding a level in the sump. If too much has been put in, wait a minute or two before replacing the plug. Too much oil can, as in a gear box, cause overheating and even excess pressure which could blow an oil seal.

Sump Pumps

Both hand operated models and 12/24 volt electric models do make life easier. They have small bore flexible plastic suction tubes which will reach down to the bottom of most gear and reduction boxes from the oil topping up hole. The oil from gear and reduction boxes is usually pretty clean, not black like the engine oil, so there is little fear of sludge building up in the bottom of castings. Small hand pumps made for the Aqua-Marine Manufacturing Company, available on both sides of the Atlantic, do a first class job inexpensively. Mercury oil drain pump, part number C-91-34429 is available. If there are large quantities of oil to pump out or if you have a fat wallet, the more expensive electric models do a good job, even with the thickest oil. If the same pump is used for dirty engine oil, it is as well to avoid contamination by flushing it clean before inserting it into the gear sumps.

Hydraulic Gear Boxes

These have proved popular in recent years. Although they are more complicated to make, involving pressurised oil to operate forward and reverse gear clutches, they only need small mechanical power to operate the system and can therefore be controlled by light push-pull control cables, such as those made by Morse and Mercury, to name but two of the many throughout the world. These control cables have been a boon to the boatbuilder as they are so easy to install compared to the old mechanical types and the highly expensive – and sometimes not so reliable – hydraulic control systems of the past.

Engine and Gearbox Control Cables

These are self-lubricating but have to be checked for chafe and to see that they are secured to the levers at either end. Loss of control, that may cause a boat to crash into the marina pontoon, can hang on a screw and its locking nut. Look at the cable installation to make sure that the builder has done nothing stupid such as bending the cable in a curve of less than a few inches, when the minimum recommended by the cable maker is a couple of feet; fierce bending is a sure way to damage the cable and shorten its life. The outer cable covering is of plastic and although quite strong, it will chafe. Gentle support with clips can help, but do make sure they do not clamp the cable tight. Controls exposed in the open cockpit or flying bridge will obviously need more attention than those kept out of the weather. An annual greasing of levers and cable terminations should suffice and the manufacturer will advise on the right lubricants.

Oil cooling may be incorporated in the transmission system and must be looked after like the main engine cooling system. Heat exchangers need the same flushing, and rubber pipe joints the same renewal, correct clipping and winterising to avoid frost damage. If the box has its own oil filter, this should be renewed at the specified interval and certainly at the end of the season, when the box must be drained of its dirty oil and new put in for the winter; try to have the gear box oil warm before it is drained, and check the condition of the old oil as you take it out. If all is well, it should look much the same as when you put it in, but if it looks even a little emulsified, suspect that water is getting in somewhere. It may be from the gear box cooling system or, as on one problem box I saw, a core plug that happened to link up into the gear box bell housing was causing unwanted water to find its way in. The nightmare situation is to discover metal particles in the oil filter, and you should certainly have heard metal disintegrating before you have to face this shock. Greyish sludge must receive immediate professional attention.

One other problem to look out for on any type of gear box is a weeping oil seal. If tiny spots of oil are flung all over the engine space by the prop shaft, see if it has come from the rear gear box oil seal. You can do a lot to preserve this seal and the rest of the box by making sure that the shaft and gear box drive flange are lined up perfectly. There is a sloppy idea prevalent that if the engine is flexibly mounted, with flexible coupling in the final drive, there is no need to have correct alignment. You may get away with this for a short time but, in the end, misalignment affects shaft wear in the stern tube and the seal and final bearing in the gear box. It is not often realised that on most modern craft, especially those with fibreglass hulls, there is a considerable change in shape when they are waterborne compared with when on land. This is sufficient to damage permanently a straight and solid propeller shaft, and does not do a lot of good to a flexibly connected one. The perfectionist will always undo the engine from the shaft when the boat is lifted out for the winter, reconnecting it loosely just before it is put afloat, and lining up the engine to the shaft when the boat is on the water; a wooden boat can take up to a week or so to gain its wet shape. Modern flexible engine mountings have easily adjusted feet so although the job takes a little time, it is fairly easy to do.

All gear box manufacturers ask the owner to check the oil level on the dipstick each day. The right oil is that recommended by the gear box manufacturer and there can be no deviation if the box is to survive. If I seem to be stating the obvious, I have known owners to put grease into a box designed for oil. It is convenient if the gear box has exactly the same oil as the engine, as there can then be no mix up and the spare can of oil will do for topping up both. However, many gear applications require Extra Pressure or EP type gear oils. Check that you are doing the right thing by your gear box.

There are two final points I would like to make. First, all owners should know if the gear box on their boat can be trailed safely or not. Makers always specify if a gear box can be allowed to rotate freely when it is trailed through the water as the boat is under sail. With some units it is perfectly safe, while others must only be allowed to rotate for a specified length of time before the engine must be started, so that oil circulating pumps

can relubricate the gear system. It is very expensive to trail gear boxes where the manufacturer specifies that it should not be done and suggests a brake be fitted on the propeller shaft to prevent it happening. This is one more item to service; it must stay clear of the shaft when drive is engaged, and stop any rotation when the boat is under sail or on tow.

Lastly, it is worth knowing what emergency facilities for ahead drive the gear box has. Although gear box failure is rare, it is possible to lose a clutch, and in some boxes you may not be able to engage a gear to get you home; hence the reason so many boats have twin engine installations. Some excellent marine gear box manufacturers, such as Newage who make the PRM 160 and 310 marine gear boxes, have simple clutch-engaging procedures which allow their boxes to engage ahead gear if a failure occurs. True, you must only proceed at one third throttle and there are other warnings but, when all is said and done, it is better to limp to port yourself than be under the obligation of calling out the rescue services. Check what the position is with your particular gear box, and find out if the box can be run for a long period in reverse. I did hear of an owner who ruined an expensive box when, on failure of the forward gear, he proceeded stern first along a canal for some hours. The reverse gear objected and he had to buy a complete new box.

Outdrives

These have gained enormous popularity—with builders because of ease of installation and with owners by providing a reliable low maintenance means of marine propulsion. But the outdrive has some limitations when it comes to owner maintenance.

Generally speaking, when mechanical things go wrong inside, it is beyond the amateur mechanic to put them right. It seldom happens, but when it does it can be expensive. The world's leading makes, such as Volvo-Penta, Mercury and BMW, have developed their products, over the years, to a state of enviable reliability; but even the best can be abused or, through some production error, have a weakness.

Major faults will have to be dealt with by a main agent using specialised tools, which even the keenest enthusiast would find uneconomic to buy. Serious mechanical faults will also demand the services of a works trained mechanic, as the outdrive is a complex piece of precision engineering that will not respond well to a dirty workshop or bad practices. There are, for example, many bearings within the housing which do not take kindly to being gummed up with excessive sealant used in the bearing carriers. Shimming is often advised in order to take up play caused by wear in pinion gears, bearing cups and to reduce gear backlash. Precise measuring instruments are necessary to determine the correct measurements of

wear, and a really good mechanic will make a record of the shims he uses, together with their location. Gears must always be mounted to the correct feeler gauge reading, and with specified backlash tolerances to avoid noisy performance and premature failure. Having frightened you with the complexity of the outdrive, may I now assure you that there are jobs that you can and must do to gain the maximum reliable service from the unit. They are as follows:

(a) Check lubrication points and use a grease gun or topping up procedures as recommended in the handbook. Because an outdrive immersed in the water may have a grease point difficult to reach without going overboard in the tender, this is no excuse for ignoring its servicing. The servicing intervals are recommended by the maker. Two types of lubrication may be employed: immersion of the gears in the oil, or by means of an oil pressure pump giving forced lubrication. Fig 57 shows the BMW outdrive unit and the parts that will need attention, and which may be found on other makes of outdrive.

Observe the old oil when draining down. If it is milky it is becoming contaminated—emulsified with water. Suspect a crack in a casting or a faulty oil seal. Usually it is the latter and this will need replacing by the professional.

Outdrives do have some protection if they strike an underwater object but occasionally, even with protection, damage can ensue. If greyish oil emerges – having a metallic

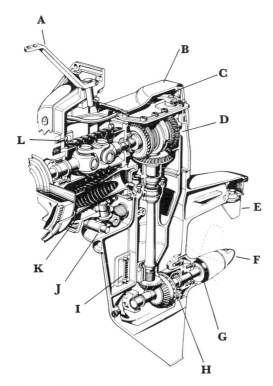

Fig 57. Servicing considerations on an outdrive leg (BMW shown).
A) Steering connections; check ease of movement and tightness of fasteners. B) Scratches must always be paint protected as corrosive attack to them is rapid. C) Outdrive oil levels need maintaining. D) Clutch for forward and reverse operation. E) Trim tab; adjust as necessary for your craft. F) Always remove propeller before running engine ashore; a cotter pin and propeller nut secures it; grease the shaft well to protect it both during the winter and when under water. G) Cathodic protection; renew. H) Outdrive gear pump; an agents job to service. I) Raw water inlet. J) Power trim/tilt hydraulic cylinders. K) Rubber exhaust bellows and (not shown) rubber water intake pipes; owner can service. L) Rubber bellows protection for drive shaft; owner can prolong life by proper use (see text) but agent should service.

appearance – it may mean a seriously worn part. Although the outdrive may be provided with an oil strainer this is definitely not a filter. It will only prevent large metallic particles from wrecking pump and gears, but will let through small particles of a disintegrating bearing. These will rapidly chew up parts in the rest of the engine and, as with any mechanical noise in an outdrive, calls for the attention of a professional mechanic.

Oil changes and grease points will usually last a whole season of about 100 hours. However the keen owner who does in excess of this will be faced with maintenance while the boat is in commission, but beaching or placing on a slipway are alternatives to expensive hauling out. For those who are keen on maximum performance, the job can be done at the same time as bottom scrubbing.

(b) Rubber connections for drive protection, exhaust outlet and cooling water intake must have regular care and inspection. The rubber concertina bellows-type sleeve which protects the main drive shaft from the engine through the transom to the head of the outdrive is, like any rubber item, vulnerable to ageing and hardening. If it cracks, it will let in water to damage the drive. Volvo Penta recommend that a new sleeve be fitted every year and that, following the instructions in their manual, it would be a twenty minute job for the amateur. BMW say that only if deterioration is noticed does this job need to be done, and even then it is the province of the professional. The one factor that will cause the sleeve to deteriorate is leaving it in the distorted position, as when on a mooring or laid with the outdrive up. Many owners leave it this way to stop fouling or, when the boat is laid up for winter, to keep the outdrive off the ground. Distortion of the bellows means permanent stretching of the underside and cramping of the top side bellows rings, which is why they fracture, especially on the underside where most stretching occurs. The bellows are usually of a large diameter and they have metal expansion rings or spiral springs on the inside to prevent sagging onto moving parts. Do ensure that you never run an outdrive when it is raised, as the spring and bellows may come into contact with the drive shaft; boats with hydraulic lift sometimes do this accidentally when, perhaps, running onto a beach. Bellows are secured to the inboard side of the engine and outdrive by means of two stainless steel hose clamps, which are also used to secure the exhaust hose and water intake hoses; they must be tight, so check them during the season. You might think that putting hose clamps on at a perfect right angle to the axis of the hose would be easy, but I was told a salutary tale by one outdrive manufacturer.

A Mediterranean owner complained that the engine on an outdrive was perfectly all right when started but, as soon as drive was engaged and the boat shot off, the engine overheated. Water pumps, thermostats and hoses – all were investigated and found in order. The agent, driven to distraction, called for a works engineer who had to travel across a continent to get to the boat. He, too, went through the engine and found, several hours later, that the water intake hose was the culprit. In the normal down position the hose drew in water but, when the driver engaged trim, because the intake hose was not on the spigot properly, though tight, a gap was opened up sufficient for air rather than water to be drawn onto the cooling system. An expensive fault.

The exhaust bellows are cooled by the ejected engine cooling water, but still operate an an elevated temperature which will age the rubber. Leakage here is not quite so serious as with the other two rubber pipes but, when renewing the others, this one can also be done as you don't want noxious fumes, or the exhaust cooling water, inside the boat.

Outdrive Power Trim and Tilt

Outdrives incorporate a tilting device so that, when the boat is being beached, the propeller can be lifted clear of obstruction and, therefore, damage. It will also allow the boat to sit on a lower trailer. Where fouling is a problem, the reduction in the immersed part of the outdrive will help although, as I have already mentioned, left in such a position it will not do the flexible rubber bellows any good. The trim devices are really a more sophisticated form of the simple tilt, the degrees of trim angle being adjustable yet positive in holding the leg steady at the maximum power transmitted through the propeller to the hull. Thus, optimum trim angles are available to cope with weight being carried and wave conditions. This enables you to save fuel by producing the maximum speed at a constant throttle opening. Displacing water unnecessarily, whether by porpoising or digging in the stern, is wasteful of fuel.

Having said what an excellent job they do, there are some points which need watching. They add to the complexity and therefore the cost of the outdrive. The simplest units have a manual lift and pins inserted to achieve optimum trim. These have next to nothing to go wrong with them, although they are unsophisticated and cannot be adjusted like the power operated ones.

Power trim engineering usually involves a small electric motor driving a hydraulic pump, which is operated from the steering position, where trim indicator and switch are to hand. There are slight variations in design. With the Volvo Penta 280 unit, for example, there are four trim positions plus beach, while the BMW is infinitely variable for any speed and load. One typical item that you should make yourself familiar with is operating procedure for, while the former outdrive is safe to operate at full throttle when in the beach (shallow water) position, the BMW must only be used at engine idling speed or damage will result. Again, I emphasise the importance of the owner's handbook.

Unfortunately, the work that you can do on power tilt/trim devices is limited to that in the handbook and, even for the professional, a number of components used are strictly throw-away-and-replace items–especially the electric motors and the hydraulic pumps. Before you buy an outdrive, it is worth checking the cost of this kind of engineering, and weighing this against the cost of repairable units.

Checking that switches do not wallow in water and that electrical leads are in good condition and properly attached, will help the electrical side. You may also be able to check the oil level in the hydraulic system and see that hose pipes and joints are in good condition and not leaking. Hydraulics of any kind do not like dirt, dirty oil or air in the system. Drastic loss of oil level in the hydraulic reservoir demands immediate investigation. Bleeding the system should not be beyond the skill of the amateur provided the manufacturer gives clear instructions. I have my reservations, however, because one outdrive manufacturer tells his customers that the oil level needs checking every fourteen days without specifying the oil or where to check it! Electro-hydraulic systems have to be protected with devices which allow the outdrive leg to hit an underwater obstacle without damage or, at most, a minimum of damage and which will allow the leg to lock automatically when the drive is put into reverse. The importance of correct functioning is obvious but for the most part you will only be able to monitor the functions and call in the expert if things are not working correctly.

Corrosion Protection and Painting

Outdrive units are cast in marine grade light alloys that must have cathodic and paint protection, especially when operating in polluted fresh or sea water. These alloys are particularly vulnerable to electrolytic corrosion and copper is a deadly enemy; it is important, therefore, never to use a copper based anti-fouling anywhere near an outdrive unit.

Fig 58 shows the BMW outdrive and indicates the position of two sacrificial zinc anodes which give protection to the large cathodic area of the alloy outdrive itself. You will find similar anodes on Volvo Penta and, indeed, on most outboard engines where a great deal of light alloy is used. The anodes, rather than the main engine, are meant to be eaten away by electrolysis and must *never* be painted. Usually, a new pair of anodes each season will give protection, but there are highly polluted waters and special circumstances where they will wear away in a much shorter time. Be vigilant and replace them when they are well eaten. There are electronic systems, worked from a battery, which give cathodic protection to craft and the Mercathode anti-corrosion system made by Mercury Marine is advised where cathodic protection is a major problem–say, for out-

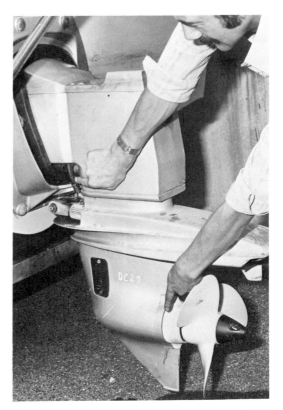

Fig 58. Cathodic protection points on the BMW outdrive.

drive units installed in metal hulls. It is important to seek professional advice but, basically, these units provide a reverse current which cancels out that generated by the dissimilar metals of the outdrive and hull; it is essential that the strict balance between these two potentials is kept. A monitor made by the same company does this effectively, and Mercury agents are advised to give the system a full test when visiting a customer's boat. It is worth remembering that an active anode will not foul, so suspect failure if you see weed or other fouling building up on it.

Painting is not, in the case of outdrive units, a cosmetic operation. It is important to avoid exposure of small areas of bare alloy to electrolytic attack. Although cathodic protection as a whole will give protection, if small areas of alloy are exposed, they can act as small anodes as well, and will corrode.

Inevitably, during the season, the outdrive will have bits of paint knocked off it–perhaps by floating debris, beaching badly or some other unwanted encounter. At lay up time when the boat comes out of the water patch up the damage; don't wait until spring. Specific information is given in Chapter 8, but basically all you need to do is to clean out the damaged area, then etch, prime, fill and finish with a preparation appropriate to the alloy.

Sail Drive Units

Where small engines are required, such as in motor launches up to twenty feet or so, or in sailing boats around the twenty-five feet mark where only auxiliary power is necessary, sail drive units are an attractive proposition.

These may be likened to an outboard unit with a power head driving the propeller through a vertical drive shaft, and a lower right angle gear to provide the horizontal drive to the propeller. Modern yachts, with their large gap between fin keel and rudder, are doing much to encourage the installation of sail drive units. Ease of installation, low drag and low weight are in their favour, though I feel that one disadvantage is the large hole needed in the bottom of the boat for the drive shaft. The makers overcome the sealing problem in a number of ways. Volvo units (Fig 59) have a really hefty 'O' ring which absorbs vibration, allowing a flexibly mounted power head to be isolated from the main structure of the boat. Renewal of this seal is straightforward and must be done every five years. It must be inspected every season and kept perfectly clean of debris, oil and solvents which could drastically shorten its life. The power head from Honda has been skilfully marinised by Volvo to produce a reliable twin cylinder overhead camshaft four stroke unit. Ignition is by breakerless capacitor discharge system which is not to be tampered with by the amateur. The rest of the engine is well within the scope of simple maintenance procedures, following the instructions in the handbook.

One point not in the handbook, is that on some early engines it was not possible to get the raw water impeller pump to suck up cooling water from a bucket when the unit was laid up on dry land. This was because

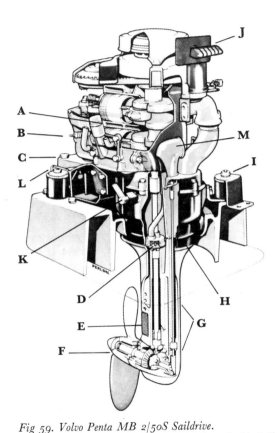

Fig 59. Volvo Penta MB 2/50S Saildrive.
A) Oil filler and dipstick. Oil removed through this hole
with an oil pump. B) Spark plug; clean and gap to
0·023″-0·027″. C) Silencer with water lock; empty with
a pump at laying up time. D) Raw water cooling im-
peller pump; see special note. E) Cathodic protection.
Renew as necessary. F) Remove propeller assembly when
running engine on land and use rust inhibiting oil on shaft
for winter. G) Leg oil level and drainage screws. H) Rubber
seal; renew as necessary but certainly every five years.
I) Flexible mountings; keep clean and free of oil.
J) Recoil starter. K) Raw water cooling cock. L) Travel
limiting device. M) Air intake; seal off for the winter.

the pump was mounted so high up the drive shaft, and it was modified on later motors. If you have an old model, Volvo will do a small modification to ease flushing procedures.

Any small engine using a recoil starter must have the starter cord inspected for chafe, and it should be renewed immediately if there is any wear. Ensure that, at the beginning of the pull when you start the engine, the pawl fully engages a ratchet before you apply full strength or a fast pull, or the mechanism may be damaged. The pull should be as direct as possible, so that the chafe on the cord as it passes through its guide hole is kept to a minimum. Never pull the cord to its extremity. When renewing it, do use the engine manufacturer's notes, as the recoil spring that pulls the cord back into the housing, retaining it there until you pull again, can be the devil to tame if it escapes. Take care because your eyes could be damaged. It is advisable to wear goggles or cover the unit with a sheet of clear plastic so that you can still see what you are doing. Sometimes the end of the spring is hooked so that it engages in a slot to keep its wound tension. This is a place where it may snap from fatigue. It is never worth putting back even partially worn small inexpensive parts, as they might let you down at a most inconvenient moment, so order a few spares and take your time to do a proper job. Both Seagull and Volvo sail drives have an accessible emergency cord start under the easily removed recoil starter unit.

Latterly, British Seagull enthusiasts have been delighted to welcome a small outdrive unit from that company. This is the first inboard power unit the company has produced but, as the power head is based on the well proven Seagull 110 outboard unit, there are no fears about reliability. Seagull acknowledge that the weakness in any saildrive is the hole in the hull, and they have made their units with solid plate mounting which is glassed into the hull. There is no seal to worry about, although I would suspect that there would be some vibration because of the method employed. At least the engine cannot be stolen from the boat like an outboard and, with the protection of a small skeg, it should provide a simple inexpensive power unit for the small open launch and sailing boat. The company has always run an excellent low cost spares service. The unit uses WIPAC capacitor discharge (CD) ignition for the two-stroke power head which has a quick bleed-down, so avoiding electrical shock when dismantling. It is worth repeating that some CD systems are dangerous to dismantle without full working knowledge. It is not possible for the amateur to test the circuit on the WIPAC unit, so it is a replaceable one. Although very reliable, it would be prudent for CD ignition owners to carry spares for small engines such as this Seagull.

One feature of sail drive units is that oil levels and condition in the lower drive unit must be checked. Both Seagull SD110 and the Volvo MB2/50S use EP 90 gear oil, and the service must be carried out when the boat is ashore or the unit is clear of the water when the boat lies alongside a quay or on a grid. Normally this operation will be avoided as Volvo give 200 hours as the service period and Seagull once a year. However, the prudent owner may like to check during the season just to ensure that all is well, especially before a long cruise. Seagull say it is quite normal for a little water to enter the box, and this should do no harm provided it is not left in all winter. After filling or checking, see that the filler and drain plugs are secured.

eight
Sprucing Up Engines

In Victorian England, cleanliness was said to be next to godliness and a slovenly person was assumed to have a slovenly mind. Times and beliefs have changed, but I have a sneaking feeling that what happens in the engine compartment of the average boat is a pretty clear indicator of the care and sympathy the owner shows his engines – and, indeed, the kind of owner he is. My private nightmare is to raise an engine hatch and see a squalid mess of rusting iron, chewed up nuts, flaking paintwork oozing a mixture of bilge water and thick oil, and pipework growing salt crystals on weeping joints, like some mechanical fungus.

The case I plead for keeping engines clean, is that it is so much easier to see what is going wrong against a background of fresh paint. For this reason, many enthusiasts treat the engine with a heat resistant white paint so that dirty oil or water instantly reveals itself. There is a good sense in this, though I do draw the line myself at having lots of shining brass and chrome; I can admire it, but I hate polishing it. Brightwork in this sense defeats its purpose by becoming a tedious chore as sea water or a salt atmosphere sets up corrosion. It is also as well to remember that sometimes when manufacturers marinise parts, the only difference between land and marine units is a coat of protective paint. For example, you may find that a pump with a cast alloy body will buff up and look great on a hot rod or custom car. If you buff up the same unit and put it on a

seagoing boat engine, there will be pits eaten into it within weeks. Paint protects as well as beautifies. It is as functional on the engine as the anti-fouling on the bottom of the boat, and it is much more pleasant for you or your mechanic to work on a clean engine. If you are paying a mechanic to work on the engine, it is best to pay for his skill in repairs rather than his skill in cleaning them to find the parts that have gone wrong.

Degreasing

Most garages stock powerful degreasing solutions which are usually applied all over the parts to be cleaned and then washed off with fresh water; if the solution is applied too liberally, it can cause chemical or water damage. Most solutions form an emulsion with the grease which is then flushed away with water. If you are concerned about water reaching vulnerable parts, it is a good idea to seal them off with masking tape and old plastic bags or polythene sheet. Watch out for the vulnerable parts of starter motors or generators, air intakes, oil and dipstick holes, distributors and any electrical wiring; none of them will take kindly to being soaked with these things. A big old shaving brush or short mop-headed brush is good for applying degreasing solution but really stubborn thick grease may require stiffer bristles. A rag soaked in solution will clean lightly contaminated areas and a ten minute soak will generally be sufficient with most cleaners. For small but important items such as electronic ignition systems, distributors and the like, there are excellent aerosol degreasing sprays which are specially formulated so that they do not harm the component materials, especially rubber and plastic.

A lint-free cloth should be used to dry the engine off after washing. Check the rag to see you have got all the grease and dirt off and, if not, treat small areas again.

Surface Preparation

It is as important to keep abrasive dust and flakes of paint out of the engine as it is to keep the water out. Before beginning mechanical abrasion of corroded surfaces, seal off air intake, carburettor or anything else that could be damaged.

There are many forms of rotary stripper on the market but the biggest problem is to find those that are compact enough to get into the confined spaces around the engine. This is less of a problem during a major overhaul when large auxiliary items are moved from the main body of the engine. Repainting items such as the generator may be done by the service agent when he has completed an overhaul, but it is best to get mechanical work over with before the cosmetic treatment. Abrasive particles from emery cloths, sanding discs or abrasive wheels will destroy finely machined parts.

Table 7 shows the comparative grades of abrasive products and, although you will be concerned mainly with the first four products, I have included garnet and glasspapers so that you will know what to ask for when you go to the chandlers. Although we are used to abrasives in sheet form, the same grades are available from firms dealing in industrial abrasives, such as discs for drill attachments, best sanders and small hand grinders.

The Wolf Grinderette (Fig 60) is a compact tool that can be used one handed to grind off serious corrosion and, with care, the fine grit papers available will achieve a good finish on castings to take paint undercoats superbly. The four inch diameter flexibly mounted discs soon clean up all but the most awkward corners of the engine, but care is needed to ensure that you do not remove too much metal.

Although the do-it-yourself trade will supply your needs, I have found it much less expensive in the long run to buy direct from industrial abrasive suppliers for, although minimum orders may be a bar, some obliging firms not only carry a good range but will advise, and are able to supply small quantities. Recently, flap wheels (Fig 60) that can be placed in the chuck of the electric drill have proved popular for fine finishing of metals. 40 to 60 grit flap wheels are good for removing serious pitting in the surface of metals, while a 180 grit wheel will give a good general finish to surfaces that are to be painted.

Grit Size	Aluminium Oxide	Silicon Carbide	Emery Cloth	Emery Paper	Garnet	Glass
1,200	—	1,200	—	4/0	—	—
1,000	—	1,000	—	—	—	—
800	—	800	—	3/0	—	—
600	—	600	—	2/0	—	—
500	—	500	—	—	—	—
400	400	400	—	0	—	—
360	—	360	—	—	—	—
350	—	—	—	1F	—	—
320	320	320	—	1	9/0	—
280	280	280	00 (280 & finer)	1M	8/0	—
240	240	240	—	—	7/0	00 (Flour) 240 & finer
220	220	220	0	—	6/0	—
200	—	—	—	—	—	0
180	180	180	—	1G	5/0	—
150	150	150	FF	2	4/0	1
120	120	120	F	3	3/0	$1\frac{1}{2}$
100	100	100	1	—	2/0	F2
80	80	80	$1\frac{1}{2}$	—	0	—
70	—	—	—	—	—	M2
60	60	60	2	—	$\frac{1}{2}$	—
50	50	50	$2\frac{1}{2}$	—	1	—
40	40	40	3	—	$1\frac{1}{2}$	S2
36	36	36	—	—	2	$2\frac{1}{2}$
30	30	30	—	—	$2\frac{1}{2}$	3
24	24	24	—	—	3	—
20	20	20	—	—	—	—
16	16	16	—	—	—	—

Table 7. Comparative grading chart for abrasives

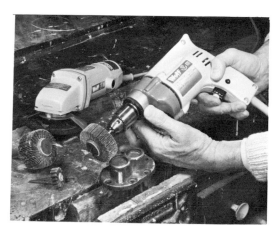

Fig 60. Wolf Grinderette and Model 3950 Drill being used to clean up small engine parts with abrasive discs and flap wheels.

Finishing

Ferrous Metal. The most perfect abrasive action is not going to strip every pock mark out of the surface of badly rusted steel or iron; you should ensure that these are neutralised. If you simply repaint them, the rust starts eating out from the pits, and will quickly lift off a new coat of paint and spread outwards. Two products I would recommend are from the Jenolite Division of Duckhams Oils, based on phosphate coatings, and those of the Owatrol Company (known as Penetol in USA) of Oslo, Norway (UK agent Geedon Marine). Both methods are simple to use, though Jenolite have such a wide product

range available that you would do well to write for literature from the address given in Appendix C. Provided the main areas of rust have been well treated, the phosphating products work well on steel, light alloys, copper and brass. With ferrous metals residual rust left after mechanical treatment is converted to inert iron phosphate.

A steel primer coat should be applied when the treated surface is thoroughly dry, the time for this being stated in the literature; this is followed with the paint system, usually a primer and a couple of top coats. One of the main points in good painting is to put on coatings at the right intervals. If undercoats harden too much, the top coat will not achieve its fullest possible adhesion properties. Rubbing down between coats aids mechanical adhesion of the following coats.

Light Alloys. After initial cleaning and preparation, light alloys in engines, gearboxes and, more especially, outdrive units, must be etch-primed. These primers are sometimes known as wash primers as they are thin solutions of phosphoric acid which create small cavities in the metal surface acting as a chemical reagent. With light alloy outdrive units, the manufacturers emphasise the importance of keeping the protective coatings intact, and usually offer touch-up paint for the purpose. However, I can thoroughly recommend the modern high performance two-can epoxy and polyurethane finishes. They need care and are expensive, but they are extremely hard wearing, withstanding knocks and abrasion far better than conventional coatings, and are more resistant to penetration by water. After etch-priming, an epoxy primer is used, followed by either the two can epoxy or two can polyurethane finishes. One precaution to take with any paint, and the high performance ones in particular, is to check compatibility with existing coatings; try out a little of the new paint's thinner on a small area to see if there is a reaction. If there is it means cleaning off all the old paint right down to the metal, a tiresome job given the intricate shapes of an engine or outdrive casting. However, if there is no reaction, you can proceed with the better paint system.

appendix A
Fault Finding Chart

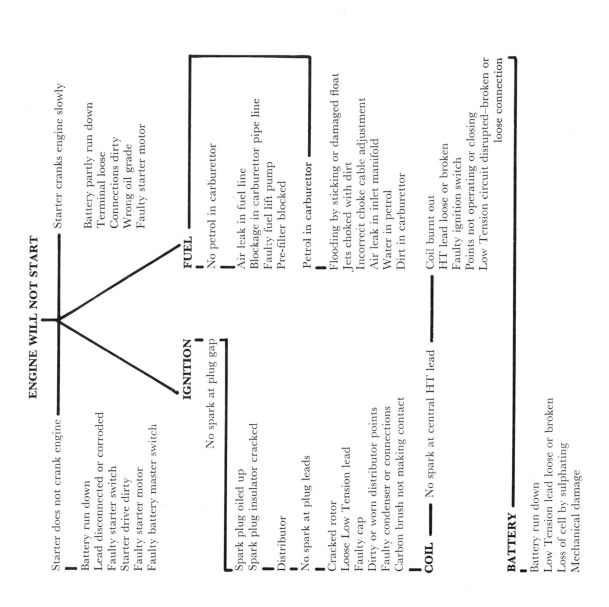

	IGNITION	CARBURETTOR	MECHANICAL
ENGINE MISFIRES	HT leads to spark plugs shorting Incorrect spark plug gap Cracked spark plug insulator Battery connection to coil loose Faulty or damp distributor cap	Water in carburettor Fuel line partially choked Fuel lift pump pressure low Fuel pump or pre-filter choked Needle valve dirty or damaged	Valves sticking Valves burnt Valve spring broken Incorrect valve clearance
ENGINE STARTS AND STOPS	Low Tension connection loose Faulty switch contact Dirty contact points	Air leaks Fuel line blocked Water in fuel Needle valve sticking/flooding, or no fuel Fuel pump faulty	
ENGINE DOES NOT GIVE FULL POWER	Fuel supply faulty Air leaks in inlet manifold Jet partly choked	Ignition retarded HT lead shorting Valve sticking or wrongly adjusted Valve burnt or badly seated Valve spring broken Wrong or faulty distributor cap	
ENGINE RUNS ON FULL THROTTLE ONLY	Slow running jet blocked Slow running adjustment screw out of adjustment	Valve sticking Valve burnt Valve spring broken	
ENGINE RUNS IMPERFECTLY	Weak mixture Fuel feed faulty Inlet valves not closing Ignition timing incorrect Carburettor flooding		
ENGINE KNOCKS OR RUNS ON			Timing too far advanced Excessive carbon deposit Loose bearing or piston Plug leads crossed Cylinder head gasket blown

appendix B
Conversion Factors for Measures and Weights

To convert	into	multiply by
C°	F°	$(t°C \times 1·8) + 32$
Centimetres	Inches	0·3937
Cubic Centimetres	Cubic Inches	0·061
Cubic Inches	Cubic Centimetres	16·39
F°	C°	$(t°F - 32) \div 1·8$
Feet	Metres	0·305
Gallons	Litres	4·546
Grams	Ounces	0·03527
Imperial Gallons	US Gallons	1·205
Inches	Centimetres	2·5399
Inches	Millimetres	25·4
Kilograms	lbs	2·205
Kilograms	Ounces	35·27
Litres	Cubic Inches	61·0
Litres	Gallon	0·220
Litres	Pints	1·76
Metres	Feet	3·28
Metres	Inches	39·37
Metres per min	Feet per sec	0·0547
Miles per hour	Knots	0·868
Millimetres	Inches	0·039
Ounces	Grams	28·35
Pints	Litres	0·568
US Gallons	Imperial Gallons	0·830
Torque		
lb/ft	Kg/m	0·138
Kg/m	lb/ft	7·233
lb/inch	N-m (Newton-metres)	0·1129
lb/ft	N-m	1·3558
Power		
Horsepower	Kilowatts (kw)	0·746
Fuel Performance		
lb/hp/hr	gram/hp/hr	447
gram/hp/hr	lb/hp/hr	0·002237
US Pint/hp/hr	cc/hp/hr	467

Specific Gravity of Petrol = 0·74

Many products mentioned in this book have been used by the author. There are undoubtedly many equivalents but I hope this information, given in good faith, will help to make life a little easier for the reader.

Product	UK Manufacturer or Agent	USA Manufacturer or Agent
Mechanic's Tools	Neill Tools Ltd., Napier St., Sheffield S11 8HB	Neill Tools Inc., 417 Woodmont Rd., Milford, Conn. 06460, USA
Magnetos Ignition Components	WIPAC Group, Buckingham, Bucks, UK	
AC-Delco Oil Filters Ignition Components Fuel Pumps Freedom Battery	AC-Delco Europe, General Motors Ltd., P.O. Box 63, London NW9 0EH	General Motors
Heli-Coil	Armstrong Fastenings Ltd., Gibson Lane, Melton, North Ferriby, North Humberside	Mercury Marine Agents
Sump Oil Pumps	Aqua-Marine Mfg (UK) Ltd., 381 Shirley Road, Southampton SO9 1HB	Aqua-Marine Inc., 2445 Michigan Ave., Detroit Michigan 48216, USA
Loctite Jointing Compounds & Sealants	Loctite (UK) Ltd., Watchmead, Welwyn Garden City, Herts AL7 1JB	Loctite Corporation, 705 North Mountain Road, Newington, Connecticut, USA
Wolf Electric Tools	Wolf Electric Tools Ltd., Hangar Lane, London W5 1DS	
Jenolite Degreasing & Rust Solvents	Jenolite Division of Duckhams Oils, Rusham Road, Egham, Surrey	
Rust Inhibitor Penetrol (USA) Owatrol (UK)	Geedon Marine Ltd., P.O. Box 11, Colchester Essex CO1 1AA	The Flood Company, Hudson, Ohio, USA
Rocol Sealants & Gasketting Materials	Rocol Ltd., Rocol House, Swillington, Leeds LS26 8BS	
DTD 369 Light Alloy Jointing Compound	Llewellyn Ryland Ltd., Haden Street, Birmingham 12	
Beazley Fire Valve	Rodal Developments, 19 Lancet Lane, Loose, Maidstone, Kent	

Product	UK Manufacturer or Agent	USA Manufacturer or Agent
Patay Pumps	Patay Pumps Ltd., The Ridgeway, Iver, Bucks	
LPS Electrical Rust Inhibitor/Cleaner and Lubricant	Conemead Ltd., Unit 23, Faraday Road, Rabans Lane Ind. Estate, Aylesbury, Bucks	LPS Research Unit Inc., 2050 Cotner Avenue, Los Angeles, California 90025

Index